D1254159

Conservation of Historic Stone Buildings and Monuments

Report of the
Committee on Conservation of Historic
Stone Buildings and Monuments

National Materials Advisory Board
Commission on Engineering and Technical Systems
National Research Council

NATIONAL ACADEMY PRESS
Washington, D.C. 1982

This study by the National Materials Advisory Board was conducted under Contract No. NSF PFR8015683 with the National Science Foundation. This contract was jointly funded by the U.S. Environmental Protection Agency, General Services Administration, the National Science Foundation, and the Andrew W. Mellon Foundation. This is National Materials Advisory Board publication NMAB–397.

Library of Congress Catalog Card Number 82–082101

International Standard Book Number 0–309–03275–X (paperbound)
0–309–03288–1 (hardbound)

Available from
NATIONAL ACADEMY PRESS
2101 Constitution Avenue, N.W.
Washington, D.C. 20418

Printed in the United States of America

NATIONAL MATERIALS
ADVISORY BOARD

iii

Technical Adviser
NEAL FITZSIMONS, Engineering Counsel, Kensington, Maryland

Liaison Representatives
HUGH C. MILLER, National Park Service, Washington, D.C.
ROBERT E. PHILLEO, Engineering Division, U.S. Army Corps of
  Engineers, Washington, D.C.
RICHARD A. LIVINGSTON, U.S. Environmental Protection
  Agency, Washington, D.C.
GEOFFREY J. C. FROHNSDORFF, National Bureau of Standards,
  Washington, D.C.
FREDERICK KRIMGOLD, National Science Foundation,
  Washington, D.C.
LAWRENCE E. NIEMEYER, U.S. Environmental Protection Agency,
  Research Triangle Park, North Carolina

NMAB *Staff*
STANLEY M. BARKIN, Staff Officer

FRANK E. JAUMOT, JR., General Motors Corporation, Kokomo, Indiana

PAUL J. JORGENSEN, Stanford Research Institute, Menlo Park, California

ALAN LAWLEY, Drexel University, Philadelphia, Pennsylvania

RAYMOND F. MIKESELL, University of Oregon, Eugene, Oregon

DAVID OKRENT, University of California, Los Angeles, California

R. BYRON PIPES, University of Delaware, Newark, Delaware

BRIAN M. RUSHTON, Air Products & Chemicals, Inc., Allentown, Pennsylvania

JOHN J. SCHANZ, JR., Library of Congress, Washington, D.C.

DOROTHY M. SIMON, AVCO Corporation, Greenwich, Connecticut

MICHAEL TENENBAUM, Flossmoor, Illinois

WILLIAM A. VOGELY, Pennsylvania State University, University Park, Pennsylvania

ROBERT P. WEI, Lehigh University, Bethlehem, Pennsylvania

ALBERT R. C. WESTWOOD, Martin Marietta Corporation, Baltimore, Maryland

NMAB *Staff*

K. M. ZWILSKY, Executive Director

R. V. HEMM, Executive Secretary

Preface

The work represented by this document grew out of a meeting of specialists concerned with the preservation of stone monuments held at the U.S. National Air and Space Museum in Washington, D.C., on September 28, 1978. The meeting was initiated by the Environmental Protection Agency (EPA) as the primary U.S. agency participating in the North Atlantic Treaty Organization's Committee on the Challenges of Modern Society (NATO–CCMS); it was organized by the Smithsonian Institution. The CCMS coordinator had proposed a pilot study, "The Impact of Pollution on Cultural Artifacts." However, the consensus of those assembled was that it would be both timely and essential to limit the study to the preservation of stone, because problems with deterioration of stone were a growing concern in virtually all NATO nations.

At the NATO–CCMS plenary session in May 1979, the pilot study was approved under the title "The Conservation/Restoration of Monuments." Greece was named pilot nation and France, the United States, and the Federal Republic of Germany the copilot nations. A copy of the proposal is included in this document.

The U.S. steering committee for the study proposed various research activities as part of this country's role. Some of these activities gained support from government agencies. For example, the Heritage Conservation and Recreation Service (U.S. Department of the Interior) undertook to develop a Census of Treated Monuments; EPA, the General Services Administration, and the National Park Service (also a unit of Interior) cooperated in a monitoring project at the Bowling Green Customs House in New York City; and the National Aeronautics and Space Administration studied applications of nondestructive analysis to the preservation of stone.

Among the activities proposed by the steering committee was a conference designed to achieve three goals: to summarize the state of research on stone conservation, to define research needs and priorities, and to interest scientists from many disciplines in the problems at hand. An approach to the National Science Foundation led to the suggestion that the National Academy of Sciences would be an ideal setting for such a meeting. A series of preliminary meetings was followed by formation of the Committee on Conservation of Historic Stone Buildings and Monuments of the Academy's National Materials Advisory Board (NMAB).

The committee organized a conference of scientists, preservation architects, engineers, and architectural historians interested in the problems of historic masonry structures. The conference was held February 2–4, 1981, at the Academy's headquarters in Washington. This document contains the proceedings of that conference and the committee's final report.

To provide a common background for the participants, the committee commissioned a series of papers designed to review the state of research in relevant disciplines. The coverage achieved by these papers is broad, but it is not comprehensive. Seismic effects, vibration analysis, removal of graffiti, surface chemistry, modeling of microclimates of buildings, and moisture in buildings are among the topics that were either omitted or given limited attention. In some areas— surface chemistry, microclimate modeling, and the effects of moisture—substitution or shifts of emphasis by the authors narrowed the anticipated coverage. Other subjects—seismic effects, adobe, and mud brick—were considered too complex to be accommodated readily in the format of the conference. Nevertheless, these assembled papers give the scientist a basic introduction to the many facets of historic preservation. Similarly, the scientific reviews give the preservation architect a general introduction to the relevant scientific and engineering disciplines.

At the conference the attendees participated in five discussion groups led by members of the committee. These frank discussions provided important supplemental material for the committee as it developed its final report.

The committee has supplemented the proceedings—the edited papers and the committee report—with certain other materials. These include bibliographies for those papers where references and reading lists were unavailable and a revised version of Bernard Feilden's previously published paper, "The Principles of Conservation." The committee's report reviews the state of research in the United States and proposes a national research effort.

ACKNOWLEDGMENTS

Many contributed to the work of the committee and the success of the conference. The enthusiastic support of Dr. Philip Handler, then President of the National Academy of Sciences, was crucial. He participated in the inception of the conference and made possible the consideration of what in many ways was a nontraditional project for the Academy. Mrs. Lee Kimche, as Director of the Institute of Museum Services, U.S. Department of Education, first suggested that a conference be held. Moreover, she played a vital role in bringing together and keeping together the several government agencies that sponsored the work of the committee. The 20 authors who contributed manuscripts in a timely manner under serious constraints on time provided the basic resource for the formulation of the final report.

Financial support for the conference organized by the committee was provided by the General Services Administration, the National Science Foundation, the Environmental Protection Agency, and the Andrew W. Mellon Foundation. The Institute of Museum Services, U.S. Department of Education, initiated planning of the conference and provided coordination of financial support. The British Council defrayed the travel expenses of conference speakers from the United Kingdom. The committee wishes to express its gratitude for these contributions.

The committee takes particular pleasure in acknowledging the splendid support of the NMAB staff, especially the project scientist, Dr. Stanley M. Barkin. His enthusiasm and hardworking stewardship provided the continuity essential to the work of the committee.

Norbert S. Baer
Chairman

Contents

Report of the Committee on Conservation of Historic Stone Buildings and Monuments

Historic and artistic monuments are the most visible aspects of our cultural heritage. These monuments, often of stone construction, are universally threatened by pollution, urbanization, public access, weathering, and other man-made and natural phenomena.

The factors that affect the deterioration of stone include its composition, structure, and surface conditions; interaction of the surrounding microclimate and environment with the stone; and the effectiveness of preservation methods. In some cases, actions intended to preserve, clean, or restore monuments have instead had deleterious effects.

We need new approaches to the conservation of these monuments. The problem involves such disciplines as geology, chemistry, meteorology, and civil and chemical engineering. Many scientists and engineers, however, are not aware of the details of the problem or of the various preservation treatments being employed. Greater progress could be made if more of the nation's scientific and technical skills could be applied to stone preservation and if greater effort were made to transfer existing technology to this field.

This committee has surveyed the extent of the problem, including the physical, chemical, and biological processes involved and the methods used to measure the rates of deterioration of stone structures. It has also reviewed the state of the art of preservation and restoration and the efficacies of the methods used. As the keystone of its work the committee convened a conference of scientists and preservationists to examine these and related matters.

The conference was held February 2–4, 1981, in Washington, D.C. The activities included the presentation of a number of invited review papers and exploration of assigned topics by five independent discussion groups. Each group was chaired by a scientist or engineer and a specialist in historic preservation. The cochairmen summarized the discussions at a plenary session.

The committee has analyzed the information presented at the conference, supplemented from other pertinent sources, and offers in this volume the review papers. In its report the committee gives conclusions and recommendations for research and development on materials and processes and their applications to the conservation of stone buildings and monuments.

The first part of the discussion that follows is organized around the topics assigned to the five discussion groups. This material is followed by consideration of general concerns that emerged in the deliberations of the several groups.

CHARACTERIZATION OF MASONRY MATERIALS

Dimension stone is among the most durable of materials, but the process of weathering alone assures that it is ultimately vulnerable. At present we have too little information about that vulnerability. Architects usually select a particular stone for its aesthetic qualities, with casual reference, at best, to basic data on porosity, pore size, moisture absorption, and other critical physical and chemical parameters. We need more effective exchange of information on stone, rather than the development of new information. A good example of what is needed is the booklet *The Stones of the Nation's Capital*, which was produced as an ad hoc activity at the U.S. Geological Survey. The booklet catalogs the building stones used in the monuments of the nation's capital. Cataloging the uses of stones in a particular region or in the nation as a whole would provide an accessible source of data useful in studying weathering.

According to the U.S. Bureau of Mines, there were 263 companies producing dimension stone in 426 quarries in 39 states in 1980. However, 10 leading operations accounted for 40 percent of the total value sold; only 53 quarries located in 15 states had sales of $500,000 or greater. Thus, compilation of data on the qualities of stone in active production could easily be made available to the architect or preservationist.

The U.S. Bureau of Mines constantly studies the market for dimension stone. The bureau maintains up-to-date inventories of producers and products. We have then the beginnings of a reference library that could collect samples and color photographs of stones and chronicle their use in buildings and monuments throughout the nation. Those interested in using a particular stone could go to a specific building or monument where it has been used and see what it looks like after years of exposure to the local environment.

We need information on the variability of stone formations, the variability of stone within a quarry, and the variability within the rock formation itself. A sample taken from the middle of a granite batholith is quite different texturally, mineralogically, and chemically from a sample taken from the side of the same batholith. Because most rock formations are not particularly homogeneous, one cannot simply take a sample of rock from a quarry, test it, and extrapolate the results to the entire quarry or formation.

Recommendation: The committee recommends the development of a national inventory of active quarries and available building stones. The inventory should include a reference library of stone samples. Sets of such samples should be kept by each state geological survey, state university, or similar organization, together with a data base of physical and chemical properties of each type and a listing of structures in which it has been used. Particular attention should be given to the variability of properties within each type and within each quarry. Ultimately the inventory should be expanded to include quarries of historical interest. Its existence should be publicized and its use by architects, builders, and conservators should be encouraged.

MECHANISMS OF DETERIORATION

In addition to observing weathering history in the field, we must determine if weathering can be recreated in the laboratory. The committee is not optimistic that this can be done.

We note also that natural outcrops of building stones have an environmental history quite different from that of stones placed on a building. In a historic structure in particular there is a humidity gradient from the outside in. Buildings that are air-conditioned in the summer or heated in the winter experience constantly changing humidity cycles. *In situ* outcrops are not supplied with water by capillary action and sprayed with salt to remove snow and ice. We must remember that a building is more than a small sample of stone in a laboratory. It is a dynamic system, made of a variety of materials and reacting to a range of stresses. We must understand how these individual factors relate to the whole and then deal with weathering and decay as a systematic problem.

Recommendation: The committee recommends the publication of a review of accelerated weathering tests and testing standards for dimension stone with a clear assessment of the applicability and limitations of the tests in predicting weathering. Responsibility for this publication could be determined by the interactive group proposed in the concluding recommendation in this summary.

Recommendation: The committee recommends the continued study of the National Bureau of Standards' test wall and the erection of other test walls in different climatic areas around the world where both stones and treatments could be tested. The purpose is to provide natural-weathering data to complement data obtained in accelerated weathering tests.

The committee has studied the papers in this volume that deal with mechanisms of decay and believes they indicate the state of the art fairly. Our discussions revealed many unresolved questions and the absence of a readily definable literature. There is literature on some aspects of the decay problem, but nearly all areas require further sustained study. Perhaps the most important matter is the effect of moisture as part of the freeze–thaw cycle or as the vehicle for migration of salts. A related problem concerns the structures of rocks. Stress–strain relationships are nonlinear, and microcracks not visible to the naked eye are extremely important. It is possible that microcracks might also provide access for movement of dilute solutions in rocks and so accelerate spalling phenomena associated with damage by salts.

More information must be developed on wet and dry deposition of air pollutants and on the whole spectrum of pollutants, not only sulfates and nitrates. Very little is known of air quality and microclimate in terms of the management of stone structures. Considerable difficulty is experienced in attempting to specify the range of monitoring required for buildings. It is not possible today to relate data accumulated at established air-monitoring stations to building sites that are often distant from these stations. Techniques for site monitoring by nonscientific personnel are essential.

Recommendation: The committee recommends the development of inexpensive methods and standards of measurement for monitoring air quality and meteorological events at buildings and other cultural sites. Such methods should be designed to monitor the special conditions that affect these sites, as opposed, for example, to conditions that may affect human health.

The roles of vibration caused by vehicles and of biological agents in the destruction of stone are other aspects of deterioration that need to be better understood. We need a systematic approach to the study of the decay of stone and the development of a general theory that encompasses all known natural and anthropogenic causes of such decay.

Recommendation: The committee recommends the development of a general theory of the decay of stone that includes all appropriate anthropogenic and natural phenomena. The model must include the effects of freeze–thaw cycles, wet and dry deposition of air pollutants, biological attack, and damage by salts.

DIAGNOSIS OF DETERIORATION AND EVALUATION OF CONDITION

In addition to their own weight, stone structures (including bridges) must often sustain such forces as geophysical loads from wind, temperature, and earth tremors; geotechnical loads from lateral earth pressures, subsidence, and foundation rotation and translation; gravity loads from walls, floors, and roofs; and vibrational loads from vehicular traffic, machinery, and blasting.

Structural distress resulting from these loads is usually manifested by fractures in the masonry itself, in the joints, or at structural interfaces. This distress tends to accelerate deterioration, compromise the safety of the structure, and detract from its aesthetic qualities.

Structural, diagnostic, and remedial techniques for stone structures are in a relatively crude state. Further, existing professional information is not well disseminated, and there is little significant structural research under way.

Recommendation: The committee recommends that the American Society of Civil Engineers (ASCE) in cooperation with the American Institute of Architects (AIA), the Association for Preservation Technology (APT), the National Trust for Historic Preservation (NTHP), the Society for Industrial Archaeology (SIA), and related professional and technical organizations develop and disseminate appropriate guides and standards for the diagnosis of structural distress in stone buildings, bridges, and monuments; for remedial actions to distressed structures; and for preventive actions to vulnerable structures.

The dominant factor in deterioration of stone and masonry structures is water and moisture. This point was stressed by many conference participants. It is obvious that considerable work in related areas should be transferable to historic preservation. We encourage the publication of papers on the subject in appropriate scientific and preservation journals.

Diagnosis of the presence of moisture is done qualitatively with a simple portable instrument that measures electrical conductivity at the surface of stone or masonry. A more quantitative portable instrument is the neutron moisture probe used in soil science and civil engineering. It is commercially available and could be used for nondestructive diagnostic surveys of structures. An infrared scanner has been used successfully to identify moisture-laden areas of large flat roofs (the moisture provides greater heat conduction, which is recognized by the infrared scan). The scanner costs about $10,000, but the savings effected by the diagnosis can make it cost-effective. In addition, technology used in the analysis and treatment of concrete highways is readily available for use on stone structures. A simple instrument used to measure vibration in roads could be used without modification to measure vibration in and movement of buildings.

Both the oil industry and NASA have techniques to measure density, porosity, and water saturation of rocks, to make surface elemental analyses, and possibly to identify internal stresses and fractures. These techniques could be applied to dimension stone, but currently they are far too complex and expensive for use by preservationists. Making current instruments portable, safe, and easy to use by the nonscientist would require a significant engineering effort. Moreover, the readings must be readily interpretable in terms of helping the preservationist diagnose problems and devise treatments. The committee believes that the transfer of this technology—developed at great cost by industrial and governmental research laboratories—to the preservation community is a reasonable goal.

Oil industry research laboratories are seeking to extract oil from rock by reducing surface tension. If this work succeeds, some of the resulting technology may be useful in reducing the surface tension in the pores of rock and other masonry materials to discourage capillary action and rising damp.

Recommendation: The committee recommends that methods and equipment developed by NASA and the oil industry for characterizing and modifying rock and phenomena associated with it be adapted for use by preservationists. We urge preservationists to explore the uses of existing portable equipment employed by NASA, the oil industry, and bridge and highway engineers, and we invite research designed to adapt existing but overcomplex technology to the uses of preservationists.

TREATMENTS FOR PRESERVATION AND MAINTENANCE

Preservation and maintenance cannot be separated, although the requirements of an initial treatment may differ significantly from the requirements for maintaining an existing condition. Many of the problems associated with treatment involve the unavailability of existing information. This information includes adequate documentation of prior treatments and the development of records, documentation of materials and methods used, and listings of materials suppliers. Trade associations and other organizations that deal with potentially relevant information should be made aware of the problems and approaches of historic preservation.

Recommendation: The committee recommends continued support for the development of the Census of Treated Monuments initiated by the Division of Technical Preservation Services of the National Park Service. Development of this data base would do much to improve the level of documentation of treated monuments.

Recommendation: The committee recommends that experimental treatments not be applied to registered landmarks and that adequate documentation be required as a condition for the treatment of registered landmarks.

Methods of cleaning stone structures require more study. There have been a number of successful treatments, but we are far from a rational, systematic approach. Criteria for coatings and vapor barriers—their permeability, mechanical behavior, and compatibility with the substrate—require considerable research.

The question of consolidation of stone has implications beyond, for example, simply lining pores and hindering the migration of various constituents in the overall system. After an initial treatment is applied, should it be used again, or is the once-treated system a completely different system? Questions of compatibility that arise here are not well understood.

Further, treatments themselves are not permanent, and the conservator must be prepared to take continuing responsibility for the stone's condition. We recognize that those responsible for initial treatments are often not those responsible for subsequent maintenance; therefore, maintenance personnel must be made aware of the principles of historic preservation.

It is important to document methods for the care and maintenance of stone to provide a basis for the design of new buildings. There is also a need to document the mechanisms of deterioration through the study of buildings that are being demolished.

Standards for materials, quality control, recommended practices, testing laboratories, and performance are required. Organizations such as the International Union of Testing and Research Laboratories for Materials and Structures (RILEM), the International Standards Organization (ISO), and the American Society for Testing and Materials (ASTM) certainly could assume leadership roles in this respect. For example, the Committee on Stone or the newly established Subcommittee on Building Preservation and Rehabilitation Technology in ASTM might very well be expanded to handle some of these responsibilities.

Recommendation: The committee recommends that ASTM, ISO, and RILEM take the lead in developing and adopting appropriate standards for performance, materials, laboratory and field testing, and practice in stone conservation. These bodies would work cooperatively with the Association for Preservation Technology (APT), the American Institute of Architects (AIA), and other professional organizations concerned with historic preservation.

EVALUATION OF TREATMENT FOR PRESERVATION AND MAINTENANCE

The conference was marked by continuing skepticism over the ability of laboratory tests to evaluate treatments. A recurring theme was the importance of field data. Although we have good control in the laboratory, we have problems with scale, with reproduction of complicated ambient conditions, and with simulating 50 or more years of weathering in hours or days or weeks.

The committee recognizes that a great deal of information on evaluation of treatments already exists if it can be found. Almost all the treatments that have been proposed have, in fact, been tried somewhere. But we need information on how well these treatments have worked. The earlier recommendation that the development of a Census of Treated Monuments in this country be supported on a continuing basis thus becomes even more important. Such a census would include careful documentation of all treatments, including, where possible, those applied in the past. We must, however, remember that reports on treatments are much more likely to detail successes than failures, whereas information on failures is at least as important as information on successes.

There is also a need for improved, nondestructive methods of evaluating the progress of treatments in the field. Photography has been used successfully to evaluate treatments; the approach ranges from elaborate, rectified photographs to very simple photography. Another evaluation method is ground-penetration radar. In its normal form, such radar has a large range and not very good resolution. But by going to higher frequencies—up to 900 MHz—good resolution is possible up to 3 or 4 ft, which is probably adequate for the purpose. While it is rather difficult to interpret the results of ground-penetration radar, it may be used as a diagnostic tool to measure the effectiveness of treatment.

Ultimately laboratory tests are required because so many more things can be tried in the laboratory than can be afforded in the field. Laboratory results must not be overinterpreted, yet we must put some reliance in them. The conference attendees agreed that if we understand the mechanisms of decay involved and can characterize the material we are working with, we can understand why it fails and can select means of eliminating that failure.

RILEM has proposed a number of nondestructive tests that may be applicable. Volumes IV and V of the RILEM-UNESCO meeting in 1978 include a large number of tests that have been proposed for use in the field. Among them are several kinds of evaluations of strength and of various aspects of porosity, permeability, and conductivity in rock. There has been no attempt as yet to set limits on these tests, but a beginning has been made. This effort should lead to a period of standardization.

Evaluation of treatment should be quantitative for a single discrete element or operation. However, there can and should be evaluations of entire systems that address elements such as costs, time, management, aesthetics, and acceptability to the clients. Perhaps we will see the development of a new generation of specialists in the administration of cultural property who will oversee activities down to the level of building management and maintenance. These would be persons who have had thorough basic training in areas such as architecture, engineering, and history and who identify personally with the properties entrusted to them. We might look to such specialists for subjective evaluation.

Recommendation: The committee recommends the expansion of the RILEM effort to refine nondestructive tests for evaluating treatments in the field. Research is needed to develop a better understanding of mechanisms of failure of treatments.

EDUCATIONAL AND INFORMATION SERVICES

Cooperative efforts between scientists and engineers on the one hand and historic preservationists and conservators on the other are limited by lack of a common vocabulary. This flaw could be remedied by the two groups' meeting together more frequently, considering each other's problems and possible solutions to them, and presenting technical concepts in terms both groups can understand.

Recommendation: The committee recommends that organizations such as the National Trust for Historic Preservation, the Association for Preservation Technology, and the U.S. National Committee of the International Council on Monuments and Sites organize scientific advisory committees within their structures to bring the scientific and conservation communities closer together.

We also need greater awareness among graduate advisers and students of the interesting scientific problems posed by historic preservation.

Recommendation: The committee recommends that universities with graduate programs in historic preservation develop strong curricula in the technology of historic buildings.

FUNDING FOR RESEARCH

The committee considers it inappropriate to suggest specific levels of funding required for the tasks outlined above. At the same time, however, we strongly urge federal agencies to recognize that conservation and preservation of monuments is eminently important and a legitimate area for research funding. A major obstacle to funding is the interdisciplinary nature of the field. Senior investigators who readily obtain support for research in a single discipline commonly encounter great difficulty in obtaining funding for interdisciplinary work. To obtain funding a researcher must restrict his proposal to the narrowly defined interests of an existing funding program within a single agency. In a field so obviously dependent on a multidisciplinary approach, where the review mechanisms of few agencies are prepared to respond favorably, it is virtually impossible to sustain a research effort.

Recommendation: The committee recommends the establishment of an interagency task force with representatives from such funding agencies as the National Park Service, the Environmental Protection Agency, the National Science Foundation, and the National Endowment for the Humanities to address the problem of funding interdisciplinary research and to establish a coordinated policy for the review and funding of research on the conservation of buildings and monuments.

To supplement funding from the public sector, those engaged in research will have to broaden their base of support to include corporate foundations and trade institutes. Where appropriate, the preservation community should explore sharing specialized equipment and support personnel.

Recommendation: The committee recommends that the preservation community seek research support in the private sector from such trade organizations as the Masonry Institute and the Brick Institute and from foundations such as the Mobil Foundation and the Alcoa Foundation.

POLICY AND PLANNING

There is at present no laboratory in the United States dedicated to the preservation of stone, nor is there a visible potential for the establishment of such a laboratory. (As a historical note, the National Science Foundation operated a Prevention of Deterioration Center from 1942 to 1964.)

The preceding recommendations call for a range of actions by government agencies, industry, universities, professional organizations, standards organizations, and individuals concerned with historic preservation. Taken together, these recommendations form a first outline for a coordinated national plan.

We need some form of organizational mechanism to bring together the existing disparate groups so that they can work cohesively to promote, coordinate, and implement activities in the conservation of historic stone buildings and monuments. No candidate for this role has emerged from the broad array of existing organizations.

Recommendation: The committee recommends that representatives from scientific and engineering professional organizations meet with representatives of historic preservation and conservation organizations to develop an intersociety mechanism to coordinate preservation activities among their groups, to expand awareness of public policy, and to develop a coordinated national plan.

The Scientist's Role
in Historic Preservation
with Particular Reference
to Stone Conservation

GIORGIO TORRACA

APPLICATION OF SCIENCE TO CONSERVATION

In 1818 Sir Humphrey Davy was sent by the king of England to Naples with the task of speeding up the unrolling of charred papyrus scrolls that had been discovered more than 60 years before in a villa near Herculaneum. In the previous years a local friar had devised a complicated contraption with which he was able to separate the papyrus sheets, but with extreme slowness; only a few scrolls were opened in that period. Because it would have taken centuries at that rate to unroll the whole library, all intellectual Europe became impatient. Everybody wished to know the content of the first classical library ever discovered.

Davy tried rapid chemical means on 11 scrolls, all of which were destroyed in the process before any attempt to decipher them could be made. Some people think he was unlucky in the choice of scrolls he submitted to the experiment, others that he was dealt the wrong ones on purpose. Whatever the case, one of the first recorded attempts of scientists to meddle with the conservation of antiquities was a complete failure.

From a superficial examination of conservation history it appears that the record did not improve very much in ensuing years. Relevant

Giorgio Torraca *is Deputy Director, International Center for the Study of the Preservation and the Restoration of Cultural Property, Rome.*

cases include the application of alkali silicates (Kuhlman ca. 1830) and fluosilicates (Kessler in 1883) to stone conservation. Little by little, scientists learned to keep away from the dangerous domains of practical conservation. Instead they wandered into the safer pastures of the analysis of artifacts, from which a flourishing new branch of science (archeometry) now grows. Scientific concepts and modern materials have obviously influenced modern conservation practice, but only insofar as they have been absorbed, more or less correctly, by the conservators who tried to adapt them to their needs.

SCIENCE AND TECHNOLOGY

Some of the difficulties met in the transformation of scientific ideas into conservation processes are common to any branch of technology. In order to overcome these difficulties and to reduce the number of costly failures, industry relies on a particular class of persons of various (and frequently dubious) extraction who claim to be able to translate laboratory data into efficient processes on the production line. I shall designate them as "technologists" or, not so respectfully but more fondly, as "tinkerers."

A constant characteristic of any job involving complex techniques is that decisions must be made before a given deadline. There is never time enough to obtain all the data required to build up a scientific model of the situation. On the production line, decisions are invariably made on the basis of insufficient information. Errors are frequent, and experience is gained the hard way, by trial and error.

The technologist is accustomed to taking risks, gambling on his innate visionary gift to build models from insufficient data, on the experience obtained from previous failures, and finally on his luck. The technologist on the production line is normally the one who survived, so he is bound to be very good or very lucky.

Progress in technology may arise from the introduction of new scientific ideas from the research laboratory, but frequently it stems from innovation on the production line attempted by a tinkerer of genius who had an urgent problem to solve. It is up to technologists to apply scientific discoveries at least as often as it is up to scientists to explain why something works. In both ways progress is made.

PECULIAR ASPECTS OF CONSERVATION TECHNOLOGY

If life is difficult for the technologist on any production line, it is even more so in practical conservation. Many variables are involved in conservation problems, and some of these lie out of the field of competence of any scientist (e.g., historic and aesthetic values). Even within natural science, the disciplines concerned are so varied that the case in which a single scientist may feel competent over the entire field is rather the exception than the rule.

The immediate result of conservation processes is evaluated by a customer (historian, architect, or layman) who has little knowledge of the processes and the risks involved, but has strong views on what the result should not be. Anyway, an objective evaluation is always difficult because there are no standard procedures of quality control and no recognized acceptance tests.

The final results of conservation processes can be judged only after a long time; this means that the outcome of a prototype operation is not known when the production line starts applying a new process. Because it is so difficult to judge the result (criteria of evaluation are nonscientific and the time required is long), it is not surprising that not only the fittest but also the less fit survive among the tinkerers and that the quality of the work produced is quite variable.

As a consequence of the frequent reluctance on the part of scientists to become involved with practical conservation, the conservators seldom enjoy the services of technologists and of laboratories applying sound testing procedures. Thus conservators are frequently tempted to take over the entire sequence of operations: experiment, application, evaluation of results. Occasionally conservators mask the lack of scientific grounding of their efforts by contact with some friendly scientist who offers some amateur collaboration. The scanty and usually irrelevant results of such collaboration are proudly displayed in reports and exhibitions to guarantee the scientific level of the work done, on which they had no influence whatsoever.

ATTITUDES OF THE SCIENTIST IN CONSERVATION

A scientist who decides to enter the field of practical conservation is, therefore, confronted with a production line whose quality is dubious and difficult to evaluate. The relevant technical literature is of difficult access and, when available, it looks shallow to the scientific eye (few data, no statistics, no reaction rates, no computerizable models).

On the other hand, the scientist is strongly attracted by the domain of conservation. This is partly because the objects to be preserved are fascinating in themselves and partly because, in the multivariable processes that govern deterioration and conservation, the hints for new ideas are innumerable. Even the dumbest scientist can discover new schemes in a short time by application of the most standard concepts of his normal branch of activity.

As soon as the scientist has developed some new ideas, he is emotionally involved. He starts seeing himself as a savior for the endangered antiquities. This is a dangerous attitude, both for the antiquities involved and the scientist himself; just like the conservator, the scientist tends to take over the entire field, getting rid of all the other incompetent people. It is not infrequent to see research, development, production, and (positive) evaluation of the result carried out by the same person. This practice obviously is questionable, even if one must admit that the difficulties of collaborating with people of different backgrounds, who also are emotionally involved, may be indeed infuriating.

"Interdisciplinarity" is frequently used as a word but is seldom put into real practice. Nevertheless, effective interdisciplinary work is an absolute requirement for progress in conservation.

THE CONSERVATION TECHNOLOGIST

A consequence of what has been said above is that, in conservation practice, we frequently meet two anomalous tendencies. On one hand, the conservator tends to improve his scientific background and to do his own research and development; on the other, the specialized scientist expands his activity to cover the whole field down to the production line.

A substantial improvement in conservation practice may be brought about only by more technological experimentation and more testing, under the condition that these be carried out by people and laboratories who are specialized in this kind of activity and are not emotionally involved (that is, people who have not invented a new process nor executed the job they must evaluate). In other words the screen of technological tests should be interposed as much as possible between basic research and the actual execution of conservation work. This is occasionally done in some government laboratories, but widespread action will probably require the support of university engineering departments, which have the right experience and equipment.

To standardize testing procedures, the cooperation of a wide range of specialists within the scientific community is essential, both at the national and the international level. Professional organizations like ASTM and RILEM have an important role to play here. As the improvement of the professional level of technologists and of the standardization of testing bring substantial progress, conservation will probably become far less picturesque than it used to be, but more reliable. Considering the high value of the property involved, reliability requirements should become more and more stringent, and much work will be needed to reach an adequate level in everyday practice.

Much of the trouble in the present situation lies in the insufficient number (and level of competence) of the technologists available. The ideal conservation technologist should be a man of solid scientific background, but with enough versatility and culture to be able to understand the attitudes of all types of specialists involved in the conservation process. He should have a feel for accurate measurement, fairness in judgment, and a decision-making capability. Above all, he should never invent a new conservation process.

The ideal technologist would also understand that conservation requirements cover such a wide field that it is almost impossible to find an absence of conflicting requirements in any real problem (e.g., authenticity versus mechanical strength). He knows, therefore, that he is always wrong on some accounts when he performs an actual intervention on a piece of cultural property. On the other hand, he also knows that action may be required quickly to avoid some worse evil.

In such cases the technologist must be able to select the right deadline, one that leaves sufficient time to collect the minimum of data required but is brief enough to keep the chance of a catastrophic development to a minimum. The solutions he will choose for his problems will not pretend to be perfect, but only "least evil" choices; he will be well aware of this and conscious that the case is never closed and that dangers lie ahead.

A kind of walker on a tightrope, the conservation technologist has as a guideline the so-called principle of minimum intervention. The principle requires that he disturb as little as possible the "information" stored in the objects he must deal with, while ensuring that it is possible to preserve them and to use them for the function assigned by the present society (or that will be assigned by a future one).

This attitude is very different from that of conservators (and scientists) of the past, who tended to do too much, either because of an egotistical tendency to show their ability or in the naive hope of ensuring preservation for eternity. Minimum intervention, however, is applicable only if care and maintenance are foreseen.

The advent of the conservation technologist implies, therefore, a general change of conservation policies, shifting the emphasis from spectacular performances in restoration to periodic maintenance routines (survey, documentation, monitoring, repair, environmental protection). If and when this desirable evolution takes place, conservation technology will become as reliable as railroad engineering, probably to the regret of the lovers of adventure who are so abundant in our trade. Luckily for them, the field is so complex and apparently inexhaustible that such a stage is not likely to be reached in a short time.

THE CONSERVATOR

Do not underrate the conservator. Scientists are frequently tempted to do so when they see him tinkering with "research" ideas and using very peculiar methods. The testing procedures of the conservator are unorthodox. However, by feeling the properties of materials with very accurate instruments (his eyes and hands), he can cut his way through a multivariable problem more efficiently than the scientist, who is accustomed to proceeding by logical steps and may have trouble identifying which variable is the relevant one.

Mixtures concocted by conservators for cleaning or protection purposes have frequently proved surprisingly successful when submitted to comparative tests with other products available on the market or prepared by scientists. In conservation, also, similar materials and processes may perform quite differently in the hands of different people. The difference may be not only aesthetic (which is quite important in this field), but might involve durability.

Good eyes and hands are attributes that are slowly developed in a professional career. They are not improvised. A good scientist is not necessarily a good conservator. A flawless execution with a low-performance material may produce a better overall result than the inept application of a scientifically tested procedure.

The conservator of the future will be a noble artisan educated in scientific and technological principles and humanistic culture. His eyes and hands will be as good as they have ever been, but he will have learned to allow other disciplines to aid his work and to extract from any sort of specialists information that is useful to him. He will be tendentially lazy, basing his actions on the principle of minimum intervention. Perhaps he will lose his natural tendency to overdo because he will be well paid to look upon the objects entrusted to his continuous care and to touch them only occasionally and with the lightest possible touch. He will resemble the family doctor who is paid to keep the patient in good shape, rather than the surgeon who performs spectacular operations in desperate cases.

SCIENTIFIC SUPPORT FOR STONE CONSERVATION

If conservation is a difficult field for scientists, stone conservation is one of the worst parts of it. The number of variables is enormous (particularly for objects exposed directly to the environment), deadlines are stringent (serious damage may take place in a short time), decisions are risky (models for deterioration must be oversimplified to be of any use), criticism is likely to be heavy (the result is frequently under the eye of the community), and, finally, failure is probable (the decay factors are always at work, while maintenance is costly and normally insufficient).

Fundamental and technological research have a lot to contribute to stone conservation, however. Two divergent lines of development have appeared: One moves from details to a general scheme; the other proceeds from theory to analysis of single cases.

The literature on stone deterioration is enormous, and it is not easy to handle. The available data should be rearranged in such a way that a general theory of the deterioration of brittle, porous materials can be explicitly formulated. Obviously the first formulation of a general scheme is bound to be tentative and should be continuously updated when laboratory and field research show that it is inadequate to explain the facts.

The general theory of stone deterioration must include all the relevant mechanical, chemical, physicochemical, biological, and climatological factors and should elucidate their interplay. A theory explains facts and takes definite positions about what is relevant and what is not; it does not limit itself to a list of all possible deterioration factors.

An example may help to explain the difference between an interpretive scheme (a theory) and a list of factors. Most books and papers on stone deterioration list *thiobacilli* as a possible cause of decay, through the conversion of sulfides or sulfur dioxide to sulfuric acid. Nobody ventures, however, to state if and when this process is important with respect to other possible mechanisms. Extensive researches conducted in Venice and Prague appear, on the basis of isotopic analysis, to exclude a relevant role for *thiobacilli* in those cities, while chemical–physical (and mechanical) factors are quite adequate to explain the sulfation process that takes place on the stones. A general theory on stone deterioration should include the notion that a relevant contribution by *thiobacilli* to stone decay in polluted urban areas is improbable.

The second line of development that I mentioned above, the detailed study of single cases, may appear to contradict what has been said about the first. When actual cases of decay are studied for conservation purposes, each should be examined in the greatest possible detail, and a model of the deterioration processes relative to that case only should be built. Conservation provisions should then be made on the basis of the model.

The importance of concentrating on isolated cases when considering treatment of stone may be underlined by an example. It is well known that a water-impervious surface coating may be dangerous on a porous material because stresses may be developed behind it if water gains access to the pore system by other ways (interstitial condensation, capillary rise, or rain penetration). The danger, however, is progressively reduced when the porosity of the material is lower and amounts to nil if the material is nonporous (e.g., metals). Stones of very low porosity (e.g., marble) have been protected for centuries in Europe by application of hydrophobic coating materials, generally with success.

Any general statement about the usefulness of protective coatings in stone conservation would be meaningless. There is no single stone disease, as there is no single human disease. The great variability in properties of both the stone and the environment ensures the existence of a large number of different processes; the importance of the various controlling factors foreseen by the general theory may vary considerably from case to case.

Substantial progress in stone conservation will be assured when several diseases are isolated and methods of treatment are tailored to each of them separately. Information from the results of attempts to cure the disease will play an essential part in future progress, as happened in the progress of medical practice and science.

There is no real divergence between the general theoretical research line and the practical one tied to the study of isolated cases. The interpretive models elaborated for single diseases must be consistent with the general theory, but, conversely, the theory must account for all the particular cases. The interplay between the general requirements of the theory and the facts resulting from analysis and practical experience is the basis of all scientific progress.

STONE CONSERVATION PRACTICE

Conservation of stone exposed to the environment requires the establishment of conservation policies rather than new conservation processes. At the present state of the art, the life of stone can be prolonged for very long periods if maintenance routines are established (as is done for steel structures exposed to the atmosphere). Such a policy involves considerable cost and effort. If and when it should be implemented is a political decision more than a technical one.

There is room for improvement, however. Costs may be reduced as materials with longer life and greater ease of application are developed; better aesthetic results may be obtained. But we should forget the notion that some day the perfect, eternal treatment will be discovered. Stone is inherently unstable on the earth's surface, and so are the protective materials we can think of. Any consolidation and protection system is necessarily going to be a temporary one.

We should stop thinking of ourselves as the saviors of the ruins of the past. We should tackle a more modest task—that of improving at all levels a series of maintenance processes, a production line that is already in operation and would keep going even without us. To do so, we must go outside the bounds of our fields of competence and work in close contact with fellow scientists from different disciplines and with nonscientists of variable backgrounds. That such a collaboration is extremely difficult is the plague, but also the beauty, of professional work in conservation.

The Principles of Conservation

BERNARD M. FEILDEN

The modern principles that govern the organization and application of conservation interventions have taken centuries of philosphical, aesthetic, and technical progress to articulate. The problem of conserving architecture and the fine and decorative arts is not simple. Even in a scientific age that has developed the technology of space travel and atomic power, the solution to local environmental problems still presents a challenge to the present and the future. Only through understanding the mechanisms of decay and deterioration can conservation skills be increased to prolong the life of cultural property for future generations.

The conservation of cultural property demands wise management of resources and a good sense of proportion. Perhaps above all, it demands the desire and dedication to see that cultural property is preserved. In this sense, two familiar maxims are pertinent: "Prevention is better than cure" and "A stitch in time saves nine." Modern long-term conservation policy concentrates on fighting the causes of decay. Natural disasters such as floods and earthquakes cannot be prevented, but by forethought the damage can be greatly reduced. Industrial life cannot and should not be halted, but damage can be minimized by combating waste, uncontrolled expansion, economic exploitation, and

Bernard M. Feilden *is Director, Rome Center for Conservation,* ICCROM, *Rome.*
This essay is taken from *Introduction to Conservation of Cultural Property* (ICCROM), 1979.

pollution. Conservation is, therefore, primarily a process leading to the prolongation of the life of cultural property for its utilization now and in the future.

CONSERVATION METHODOLOGY

The conservation of cultural property constitutes a single, interprofessional discipline coordinating a range of aesthetic, historic, scientific, and technical methods. It is a rapidly developing field that, by its very nature, is a multidisciplinary activity, with experts respecting each other's contributions and combining to form an effective team.

Despite the difference in scale and extent of intervention, the underlying principles and procedural methods remain the same for the conservation of movable and immovable cultural property. There are, however, important logistical differences.

First, architectural work entails treatment of materials in an open and virtually uncontrollable environment. Whereas the museum conservator/restorer can generally rely on good environmental control to minimize further deterioration, the architectural conservator cannot. He must allow for the effects of time and weather.

Second, the scale of architectural operations is much larger, and in many cases methods used by museum conservators/restorers may be found impracticable because of the size and complexity of the architectural fabric.

Third, and again because of the size and complexity of architectural conservation, contractors, technicians, and craftsmen must actually perform the various conservation functions, while the museum conservator/restorer may do most of the treatment with his own hands. Communication and supervision, therefore, are important considerations for the architectural conservator.

Lastly, because the architectural fabric has to function as a structure, resisting its own dead weight and applied live loadings, there are further differences between the practice of architectural and museum conservation. Architectural conservation must be within the context of historic structure, which also incorporates its site, setting, and physical environment.

For both movable and immovable cultural property, the objects chosen for treatment and the degree of intervention are predicated upon the values that can be assigned to the property. These values help systematically to set priorities in scheduling interventions, as well as programming the extent and nature of the individual treatments. The assignment of values or priorities will inevitably reflect each different cultural context. For example, a small wooden domestic structure from the beginning of the nineteenth century in Australia would be considered a national landmark because it dates from the founding of the nation and because so little Australian architecture had survived from this period. In Italy, on the other hand, with its thousands of ancient monuments, a comparable structure would have a relatively low priority in the overall conservation needs of the community.

The values assigned to cultural property come under three major headings:

- *Cultural Values*: documentary value, historical value, archeological and age value, aesthetic value, architectural value, scientific value, symbolic or spiritual value, townscape value, landscape and ecological value.
- *Use Values*: functional value, economic value, social value, political value.
- *Emotional Values*: wonder, identity, continuity.

For movable objects, the question of values is generally more easily defined. However, in architectural conservation, problems often arise because the utilization of the historic building, which is economically and functionally necessary, must also respect cultural values.

The costs of conservation may have to be allocated partially to each of the above values in order to justify the total to the community. There may be conflicts between some of the values. In certain cases architectural values will predominate. In other cases artistic or historical considerations will prevail, while in yet others practical and economic considerations may modify the scope of conservation. Sound judgment based upon wide cultural preparation and mature sensitivity gives the ability to make correct value assessments.

TREATMENTS

When a treatment is being planned, the following three general concepts regarding an object or a structure's condition are considered, as exemplified by a minor seventeenth-century wooden polychromed sculpture damaged in the collapse of a church during an earthquake in Friuli, Italy:

- *Damage suffered by the object.* The sculpture suffered multiple fractures during the church's collapse. The body was broken into three pieces—the head, torso, and right leg. Both feet and arms were missing and presumed lost. The borders of the breaks were very abraded.

• *Insecurity of the object.* The wood was weakened from wormwood infestation, which had reduced the wood to frass in many areas. The wooden support and glue preparation for the polychrome were further weakened by mold growth resulting from exposure following the church's collapse. The remaining original polychrome had detached and tended to fall off. Green mold hyphae were evident, and there seemed to be dry rot as well.

• *Disfigurement suffered by the object.* Much of the remaining original paint was covered by unsightly overpaint. The entire sculpture was very dusty and in some areas was encrusted with mud.

By taking the above criteria into consideration, a judiciously selected intervention was programmed in this case. Because the sculpture had relatively minor artistic and historical value, and would remain indefinitely in storage rather than on view, treatment priorities focused on conservation methods needed only to ensure the sculpture's survival: fixation of polychrome, securing of breaks, removal of surface dirt, disinfection, fumigation, and controlled drying. Restoration measures, such as the removal of the overpaint and the reintegration of lacunae, were held for some future date.

During this treatment, and during all conservation treatments, the following standards of ethics must be rigorously followed:

• The condition of the object, and all methods and materials used during treatment, must be clearly documented.
• Historical evidence should be fully recorded; it must not be destroyed, falsified, or removed.
• Any intervention must be the minimum necessary.
• Any intervention must be governed by unswerving respect for the aesthetic, historical, and physical integrity of cultural property.

Interventions should:

• Be reversible, if technically possible.
• Not prejudice a future intervention whenever this may become necessary.
• Not hinder the possibility of later access to all evidence incorporated in the object.
• Allow the maximum amount of existing material to be retained.
• Be harmonious in color, tone, texture, form, and scale, if additions are necessary, but be less noticeable than original material, while at the same time being identifiable.

- Not be undertaken by conservators/restorers who are insufficiently trained or experienced, unless they obtain competent advice. However, it must be recognized that some problems are unique and have to be solved from first principles on a trial-and-error basis.

Preparatory Procedures

Prior to conservation interventions, preparatory operations are required:

Inventories

At the national level, conservation procedures consist first of making an inventory of all cultural property in the country. This is a major administrative task for the government. It involves establishing appropriate categories of cultural property and recording them as thoroughly as possible, graphically and descriptively. Computers and microfilm records are valuable aids.

A preliminary written study of each object or building is necessary in order to know and define it as a whole, which, in the case of architecture, includes its setting and environment. The present condition of the building or object must also be recorded. Documentation of these studies must be full and conscientious. Records and archives must be searched. In some countries, reliance may have to be placed on oral traditions, which should be recorded verbatim and included in the dossier created for each object or building.

Documentation

Complete recording is essential before, during, and after any intervention. In all works of preservation, repair, or excavation of cultural property there must always be precise documentation in the form of analytical and critical reports, illustrated with photographs and drawings. Every stage of the work of cleaning, consolidation, reassembly, and reintegration, including all materials and techniques used, must be recorded. Technical and formal features identified during the course of the work should also be included in the documentation. This record should then be placed in the archives of a public institution and made available to research workers. Finally, if the intervention can in any way serve to broaden general knowledge, a report must be published. Often in large projects it may take several years to write a scholarly report, so a preliminary report or series is desirable to inform the public and to maintain popular support.

Documentation is essential because it must be remembered that the building or work of art will outlive the individuals who perform the interventions. To ensure the maximum survival of cultural property, future conservators/restorers must know and understand what has occurred in the past.

Interventions

The intervention should be the minimum necessary. The techniques used depend upon the conditions of climate to which cultural property is likely to be subjected. These fall into three groups:

- Natural climatic and microclimatic conditions, which vary greatly and are virtually uncontrollable.
- Modified climatic conditions, such as those found in a normal building that forms an environmental spatial system with a partially self-adjusting modified climate.
- Conditions where humidity and temperature are controlled artificially to minimize dangerous variations. Ideally, the climatic control has been designed for the safety of the objects, rather than the comfort of the visitor.

Interventions practically always involve some loss of a "value" in cultural property, but are justified in order to preserve the objects for the future. Conservation involves making interventions at various scales and levels of intensity that are determined by the physical condition, the causes of deterioration, and the probable future environment of the cultural property under treatment. Each case must be considered individually and as a whole, taking all factors into account.

Always bearing in mind the final aim, principles, and rules of conservation, seven degrees of intervention can be identified. However, in any individual conservation treatment, several degrees may take place simultaneously in various parts of the whole. The seven degrees are:

- Prevention of deterioration
- Preservation
- Consolidation
- Restoration
- Rehabilitation
- Reproduction
- Reconstruction

Prevention of Deterioration

Prevention entails protecting cultural property by controlling its environment, thus keeping agents of decay and damage from becoming active. Neglect must also be prevented.

Therefore, prevention includes control of humidity, temperature, and light, as well as measures for preventing fire, arson, theft, and vandalism. In the industrial and urban environment, it includes measures for reducing atmospheric pollution, traffic vibrations, and ground subsidence from many causes, particularly abstraction of water.

Preservation

Preservation deals directly with cultural property. Its object is to keep it in the same state. Damage and destruction caused by humidity, chemical agents, and all types of pests and microorganisms must be stopped in order to preserve the object or structure.

Maintenance, cleaning schedules, good housekeeping, and good management aid preservation. Repairs must be carried out when necessary to prevent further decay and to keep cultural property in the same state. Regular inspections of cultural property are the basis of prevention. When the property is subjected to an uncontrollable environment, such inspections are the first step in preventive maintenance and repair.

Consolidation

Consolidation is the physical addition or application of adhesive or supportive materials into the actual fabric of cultural property in order to ensure its continued durability or structural integrity. In the case of immovable cultural property, consolidation may entail, for example, the injection of adhesives to secure a detached mural painting to the wall. Movable cultural property, such as weakened canvas paintings and works on paper, are often backed with new supportive materials.

With buildings, when the strength of structural elements has been so reduced that it is no longer sufficient to meet future hazards, the consolidation of the existing material is necessary, and new material may have to be added. However, the integrity of the structural system must be respected and its form preserved. No historical evidence should be destroyed. Only by first understanding how a historical building as a whole acts as a "spatial environment system" is it possible to make adjustments in favor of a new use, introduce new techniques satisfactorily, or provide a suitable environment for objects of art.

The utilization of traditional skills and materials is of essential importance, as these were employed to create the object or building. However, where traditional methods are inadequate, the conservation of cultural property may be achieved by the use of modern techniques that should be reversible, proven by experience, and applicable to the scale of the project and its climatic environment. In buildings made of perishable materials, such as wood, mud, brick, or rammed earth, traditional materials and skills should be used for the repair or restoration of worn or decayed parts.

Finally, in many cases it is wise to buy time with temporary measures in the hope that some better technique will evolve, especially if consolidation may prejudice future works of conservation.

Restoration

The object of restoration is to revive the original concept or legibility of the object. Restoration and reintegration of details and features occur frequently and are based upon respect for original material, archeological evidence, original design, and authentic documents. Replacement of missing or decayed parts must integrate harmoniously with the whole, but on close inspection must be distinguishable from the original so that the restoration does not falsify artistic or historical evidence.

Contributions from all periods must be respected. All later additions that can be considered as historical documents, rather than merely previous restorations, must be preserved. When a building includes superimposed work of different periods, revealing the underlying state can be justified only in exceptional circumstances: when the part removed is widely agreed to be of little interest, when it is certain that the material brought to light will be of great historical or archeological value, and when it is clear that its state of preservation is good enough to justify the action. Restoration also entails superficial cleaning, but with full respect for the patina of age.

Rehabilitation

The best way of preserving buildings is to keep them in use, a practice that may involve what the French call "mise en valeur," or modernization and adaptive alteration.

Adaptive reuse of buildings, such as utilizing a medieval convent in Venice to house a school and laboratory for stone conservation or turning an eighteenth-century barn into a domestic dwelling, is often the only way that historic and aesthetic values can be made economically viable. It is also often the only way that historic buildings can be brought up to contemporary standards by providing modern amenities.

Reproduction

Reproduction entails copying an extant artifact, often in order to replace some missing or decayed, generally decorative, parts to maintain its aesthetic harmony. If valuable cultural property is being damaged irretrievably or is threatened by its environment, it may have to be moved to a more suitable environment. A reproduction is thus often substituted in order to maintain the unity of a site or building. For example, Michelangelo's sculpture of David was moved from the Piazza della Signoria, Florence, into a museum to protect it from the weather. A good reproduction took its place. Similar interventions were undertaken for the sculpture of the cathedrals of Strasbourg and Wells.

Reconstruction

Reconstruction of historic buildings and historic town centers using new materials may be necessitated by disasters such as fire, earthquake, or war, but reconstructions cannot have the patina of age. As in restoration, reconstruction must be based upon accurate documentation and evidence, never upon conjecture. In the case of a work of art, a stolen panel from the Ghent Altarpiece (ca. 1432) was replaced with an exact reproduction.

The reerection of fallen stones to create an accurate and comprehensive version of the original structure is a special type of reconstruction called "anastylosis."

Moving entire buildings to new sites is another form of reconstruction that is justified only by overriding national interest. However, it entails some loss of essential cultural values and the generation of new environmental risks. The classic example is the temple complex of Abu Simbel (XIX Dynasty, Egypt), moved to prevent its inundation by the Aswan High Dam.

Some Illustrative Preservation Problems and Treatments in Washington, D.C.

J. WALTER ROTH

In conjunction with a self-guided tour of historic stone buildings and monuments in the District of Columbia a diagrammatic map was prepared, noting not only the street locations of the subjects but also the nearest stations of the Metro subway system. A brief statement about each subject describes pertinent preservation problems, treatments, and evaluations—if any. The map and descriptions do not constitute a single tour to be comprehended as a unit; rather, they comprise a limited catalog of cases from which one can select those subjects of particular interest and convenience.

The subject buildings and monuments are not limited to the U.S. Public Buildings Service inventory. The basis for selection is their potential to demonstrate a variety of problem conditions and to illustrate the effects of various treatments or the lack thereof. To the greatest extent possible, the presentation incorporates expositions of problems as perceived by the users and the stewards of these properties. Aside from its intrinsic worth as a direct statement of specific site-related masonry problems and treatments, this presentation, because of its visual nature, should serve as a stimulus to suggest similar situations—possibly even solutions.

Masonry has been the architectural material of choice since time immemorial. It has been favored for its permanency and its ability to achieve spectacular effects of size and scale, texture and color—even

J. Walter Roth *is Director, Historic Preservation Staff, Public Buildings Service, General Services Administration, Washington, D.C.*

though such effects were frequently directed toward symbolic, ornamental, or literal representations of less permanent materials, objects,
or individuals.

Masonry has been an architecturally attractive material, not only
because of its inherent aesthetic qualities but also because it gave the
greatest promise of immortality to the works of man. When natural
masonry, that is to say stone, was not available, man made it from
materials at hand, using mud, clay, cement, sand, gravel, glass, steel,
and even seashells to produce adobe, brick, terra-cotta, concrete, or
tabby. Man has transformed those imitative materials into surrogates
for an ideal stone. What would make a stone ideal? Implicit aesthetic
qualities, the potential for achieving astounding structural and architectural effects—and the magic to last forever.

To last forever is not an unusual objective of people, particularly in
their collective or national manifestations, and that objective historically has been expressed in monuments erected to symbolize or shelter
their institutions. Nations invest much of their resources in those
constructions in an effort to make them permanent as well as im-
pressive. But permanency may be perceived in a passive sense to mean
that once an edifice is erected it will be there forever and nothing more
need ever be done to it. A certain amount of obvious aging of buildings
and monuments is acceptable to most observers. Such aging is analogous to the development of character in the visage of man. Venerable,
honored institutions may be better appreciated by some in edifices
that visibly exhibit their witness to history. That is typically the case
with older nations. The brand new, instant capitals of younger nations
pose an anomaly in that regard. These nations pride themselves on
their newness, their freshness, and their orientation to the future rather
than the past.

Given the opportunity and the ability, we would build our way to
immortality. It has been tried before. In spite of the biblical lesson of
Babel, we persist in our efforts to gain infinity—if not through the
extension of space, then through the entrapment of time, the attain-
ment of eternity in our monuments. Typically, our architecture, sacred
or profane, endeavors to express stability, substance, strength, and
endurance. Even in the workaday environments we create, we strive
for longevity—without limits, if possible. If, like Le Corbusier, we see
our buildings as "machines," then our ideal is to invent the perfect
independent engine—the perpetual nonmotion machine, as it were. It
may be naive for us to wish for machines that go on forever without
attention, without maintenance, repair, or replacement. But that is

what man, the immortality seeker, has always wanted. He has especially sought this ideal in his special symbols, his monuments.

One of the inspired concepts in the program of this Conference on the Conservation of Historic Stone Buildings and Monuments is the idea of a self-conducted walking tour of masonry buildings and monuments in Washington, D.C. (see pages 47–48). I say inspired because, first, the idea did not originate with me and, second, because it translates into actual experience the content (or should I say the discontent?) of our concerns.

I would submit that most people are interested observers of the passing parade and the backdrops against which life's comedies and tragedies are played. But while most people are interested observers, they are also typically casual, as opposed to clinical, observers. When we look at the pretty young girl or the bent old man, the puffing jogger or the babbling babe, we are not making a fragmentary analysis of their skin condition or the state of their arches, their circulation or their communication systems. Most people as they walk or ride past buildings and monuments do not analyze in detail the conditions manifested or implied by separate, specific physical signs. Yet many people, depending on their sensitivity, experience, or knowledge, are inclined to make intuitive or summary evaluations of structures—evaluations that may go beyond an estimation of existing conditions and proceed from diagnosis to prognosis.

I am suggesting that we are making evaluations all the time about all sorts of things. Estimations, value judgments—call them what you will—they are the stuff of cybernetics, data for decision makers. And Washington is the mecca of decision makers. Decisions minor and major, simple and complex, early and late, are made constantly—though not necessarily with constancy—among these marble temples dedicated to faith in the processes of governance.

Washington is an especially good place to observe masonry. It is a monumental city in many ways, ranging from the sweep and scale of its baroque plan to its very name, which monumentalizes the symbolic father of the nation. The accommodation of a multitude of government agencies, national foundations and associations, institutions of higher learning, and any number of organizations having special status or seeking it has produced over the years a collection of constructions in masonry unmatched in other American cities of similar size.

In spite of the predilection of our national leaders for the classical in architecture (thus ensuring a predisposition to stone), Washington has followed trends into a variety of styles, original or eclectic, but always finding its form in masonry. The masonry itself has been highly varied, selected sometimes strictly for its architectural effect, but all too often chosen for convenience: geographic, economic, or political. As a result Washington has a great diversity of stones in its buildings and monuments—some resulting almost accidentally from ready availability or success in competition, others especially solicited to serve as ambassadors, as it were, of the states and foreign countries. For instance, the exterior of the Washington Monument demonstrates the former case with a progression of cladding marbles from Maryland, Massachusetts, and Maryland again, while the interior of the monument exemplifies the latter, containing a collection of commemorative stones from all over the United States and abroad.

Washington is a place of "hype" and "hyper." The construction of the sandstone Norman Castle of the Smithsonian Institution in 1848 was preceded by a highly professional public relations campaign, including the publication of Robert Dale Owen's *Hints on Public Architecture*—an early example of the hard sell. These days, architectural proposals are previewed, viewed, and reviewed by dozens of authorities before they are finally realized, and many evaporate in the heat and pressure of the process. For better or worse, a great amount of attention is given to every aspect of planning, funding, design, and contruction. In some cases, operation and maintenance of the edifice are also considered, but in the jargon of hype and hyper those considerations do not tend to be "up front." Yet, the promise, if not the practice, of more complete custodial care and record keeping is potentially available in Washington because it is the nation's capital, the locus of national pride and international interest.

Mention was made earlier of the observation of elements or events in the environment and how those observations can lead to diagnosis and prognosis. Mention was also made of the physical embodiment of ideals and how they may be enshrined in everlasting, ever-shining glory. Sooner or later there must come the realization that unless there is constant care and renewal, entropy will take its toll.

In a system in which everything must be paid for and everything has a price—increasingly exorbitant—philosophical ideals are frequently attenuated and sometimes compromised. The physical ideals for our buildings and monuments are even more easily conceded. Mature people know that it is possible to grow old gracefully; in fact, they learn that it is an imperative. People also learn that objects can age with grace and beauty and, in fact, with increased interest. Aesthetic and associative qualities are enhanced with the passage of time and the resultant weathering, coloration, and advent of individual physical characteristics.

As fresh building materials react and adapt to the influences of climate and immediate environment they develop personalities, as it were. They also become members of the community, asserting their intention to stay. Eventually they settle in and, if they mature without unusual stress or accident, look as though they have always been there and should remain. As the process of degradation proceeds, however, the tolerance level recedes, and the degree of acceptance of these treasured old piles diminishes in proportion to their degradation. Many people would be pleased to retain the worn, weathered building for its intrinsic symbolic value. But as maintenance and operation become increasingly difficult, physical endurance, the demands of new uses, and the price people are willing to pay for retention of original fabric all reach their limits.

To compare the aging and deterioration of buildings to that of people is not far-fetched in my estimation. Similes may be made at various stages: Thorough, sensible, environmental planning and thoughtful design detailing may be compared to prenatal care; proper adjustment and correction measures may be compared to infancy and preadolescence; initial protective treatments and routine maintenance in operations may be compared to a period of growth and early maturity. We may compare special maintenance measures to the exigencies of middle age, adaptive reuse to a period of retirement, restoration and replacement to old age, and recordation to memorialization when the decision is finally made to bid farewell to old friends.

The stewardship of significant buildings and monuments entails a variety of responsibilities, particularly in a national capital. These structures serve as the backdrop for significant events or activities. Sometimes they are even the featured players in the brief dramas that flit across our TV screens during the nightly newscasts. Those who have the prime responsibility for planning, design, construction, operation, and maintenance come under the scrutiny of the Secretary of the Interior's advisory boards, the National Capital Planning Commission, the National Commission of Fine Arts, the General Services Administration evaluation panels, the General Accounting Office, and the Office of Management and Budget, not to mention numerous components of the sponsoring and using federal agencies and the government of the District of Columbia. Eventually, the ultimate critic, the client/owner (that is to say the taxpayer) has his say—and he is not at all hesitant in voicing his opinion concerning the aesthetics and utility of the very stones of *his* properties. Frequently the opinions are irate. Sometimes they even come in the form of tort claims for injuries resulting from some simple stone problem or, even worse, from a well-intentioned corrective or preventive measure. Masonry preservation problems may be categorized in several types; their corrective or preventive treatments, being an evolving and often expedient art (or science), probably defy categorization.

STAINING AND DISCOLORATION

Probably the most frequently encountered problem is the discoloration of masonry. I am not talking about the occasional efflorescence of new brickwork or the inevitable weathering of freshly worked stone surfaces, nor am I talking about the gradual effects of air pollution or the mindless impact of animal life. I am talking about the stains that result from the transfer of matter from adjacent substances.

This kind of discoloration is very familiar, a result of runoff from dissimilar materials and dirt. The most casual observer cannot help but notice the green stains on the marble bases of bronze statues or on the facades of those buildings covered by copper roofs. The old saying that familiarity breeds contempt may not be applicable; in this case, familiarity seems to breed tolerance, acceptance, expectation. That is on an emotional level. The intellectual acceptance of a coating of verdigris on stone may be accommodated by the understanding that it is not harmful, indeed might be helpful, and at least can be visually interesting.

This sort of thinking process is abetted when one can see or imply the source: roof coverings, doors, window frames, exterior hardware, sculpture, or other architectural ornament or accessories. Staining from unseen sources, such as flashing, cramps, conduits, gutters, reinforcement, wiring, and similar devices, may puzzle the casual observer and alarm the concerned professional. To the former it is at most an instance of an unaccountable effect; to the latter it is at least an indication of imminent or eventual failure, possibly danger.

Traditional materials and designs may produce staining that we are conditioned to anticipate, even to accept. But newer architectural or artistic concepts and less familiar materials bring with them different patterns and kinds of staining and are less likely to find acceptance or understanding. For example, buildings and sculptures utilizing the new weathering steels (such as Cor-Ten or Mayari-R) can produce obvious and repulsive red-brown runoff or puddle stains. Perhaps time will provide a tolerance for and acceptance of such effects; perhaps the

FIGURE 1 "The Arts of War: Sacrifice" by Leo Friedlander (1951). Runoff from mercury-bearing gilding on this bronze statue at the northeast entrance to Arlington Memorial Bridge produces pronounced staining on the granite base.

designers will learn to manage problems inherent in the nature of those forms and materials.

WEATHERING AND POLLUTION

The effects of weathering and pollution are less direct than staining from adjacent materials, but eventually can be just as obvious and problematic. Because the effects do tend to be more gradual and general, there may be an even greater degree of tolerance and acceptance than for the isolated, spotty instances of staining.

The general effects of weathering, as opposed to the particular ravages of environmental pollution, are acceptable to many people because they carry connotations of permanency and character. There may be an inclination in some sections of management to make the objective of their maintenance programs the retention of a "brand-new" appearance, but a preventive maintenance program that would provide the required level of utility while assuring an acceptable level of physical conservation of the facility is more desirable. And that goal is obtainable without eliminating symbolic significance.

Visitors to Disneyland are impressed by the high level of cleanliness, freshness, and newness that is maintained. But Disneyland is a fantasyland that exists outside of time and outside of any context other than its own. Williamsburg, in spite of its seasonal programs and daily activities superimposed to attain "living history," is a re-creation, an extended exhibit, an artificial time-island; and, although with greater decorum, it is, like Disneyland, a place of escape. Williamsburg is also a place for the interpretation of history and culture, which brings to mind a problem faced in the management of original historical or cultural properties: If deterioration, damage, or other deficiencies require the replacement of fabric, to what extent can aesthetic appearance and historical character be compromised without losing the integrity of the experience for the user?

Limited wear or weathering may be acceptable, even construed as a positive contribution, but the more insidious and damaging influence of pollution will produce essentially negative effects in the surface appearance and, ultimately, in the structural substance. The effects of weathering and pollution may be treated after the fact by repair, restoration, or replacement. But as good practice in conservation—and in view of the pervasive nature of those effects—it makes sense to practice prevention.

Given unavoidable hostile climatic conditions and undesirable adverse environmental factors, the early (if not initial) cautious use of protective coatings might be the only realistic means of delaying eventual (if not inevitable) deterioration. Such practices require careful selection and continuing attention, analysis, and evaluation. Sensible planning, design, and specification can be the best treatments against weathering and pollution.

Planning should be comprehensive and take into account the anticipated results of various orientations and exposures, accessibility to flora and fauna, and potential impacts from existing and proposed elements in the environment. The monitoring, analysis, and correction of pollution-caused environmental conditions is a long-range process, one that may have greater value as categorical information than immediate applicability as specific data.

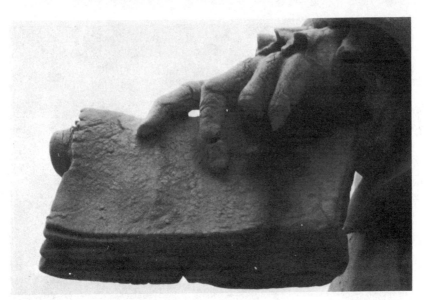

FIGURE 2   The marble statute of Benjamin Franklin by Jacques Jouvenal (1889) at 12th Street and Pennsylvania Avenue, N.W., shows weathering effects, particularly on fingers of the left hand.

IMPACTS: ANIMAL, VEGETABLE, MINERAL

Masonry structures are more vulnerable than most people realize. Although their very reason for being derives from a myth of invulnerability, they suffer in many ways. To categorize the sources of the suffering as animal, vegetable, or mineral may be more convenient than scientific. By extension these categories could conceivably accommodate such exotic damaging phenomena as earthquakes, wars, and floods.

Mineral impacts have been implied in the discussion of staining through runoff and leaching from adjacent noncompatible materials. Some mineral deposits in the form of graffiti (unauthorized) or graphics (authorized) appear on masonry; but, strictly speaking, the agent of application is animal (human).

Vegetable impacts are familiar and frequent, ranging from microcosmic mosses to clinging vines to unplanned hanging gardens growing out of convenient cracks and crevices in our buildings and monuments. They might be termed the permanent problem plant community. There are also transitory problems, such as seasonal deposits of debris and damage from the rising roots or falling limbs of trees.

FIGURE 3　Internal Revenue Service Building at 12th Street and Constitution Avenue, N.W. Architect: Louis Simon (1928-36). Birds may be the most ubiquitous of animal pests.

FIGURE 4   The Federal Building (Dept. of the Post Office Bldg.) Sculptor: A. A. Weinman? (1934?). Esthetic effects of antipest devices may be counterproductive.

Animal impacts are probably the most varied and the most damaging. Birds are among the major offenders. They manage to roost on the newer, more streamlined buildings almost as well as they do on the older, more highly ornamented ones. Their ability to adopt any kind of structure as a toilet facility seems unlimited. Their droppings are more than a nuisance; they are a known source of at least two potentially fatal diseases: histoplasmosis from a fungus and meningitis from a bacterium. The residues from their acid-laden droppings, their nests, and even their corpses all pose problems in the operation and maintenance of buildings and monuments. Many treatments have been tried: electrical (high-voltage wires, light rays), chemical ("hotfoot"

FIGURE 5   Justice Department Building at Pennsylvania Avenue and 10th Street, N.W. Architect: Borie, Medary, and Zantzinger (1934?). Expedient or unstudied maintenance methods, as in graffiti removal, may produce undesirable, long-lasting effects.

compounds, poisonous or contraceptive feeds), physical (wire mesh, spikes, slanted boards, sticky compounds), auditory (ultrasonic and sonic alarms, intermittent explosions), predatory (falcons, snipers, traps)—anything that man could devise.

Although birds, rodents and other quadrupeds, insects, even microzoa all take their toll, mankind in his enlightenment is most unkind. Man's offenses are often vicious. The normal wear and tear of traffic and toil is to be expected, but malicious mischief, vandalism, and waste in the use of our resources are shameful and cannot be excused. For that matter, there is no real basis for excusing ignorance, disregard, or neglect in the management of our resources, either. If those animals of the supposedly highest order of intelligence and trust (us) cannot govern themselves in the management of their own common property, then the problems transcend the rather limited concerns of bird droppings and freeze–thaw cycles.

CORRECTIVE MEASURES AND COMPOUNDED PROBLEMS

There is always a danger that when a problem is treated, an unanticipated reaction may result. Sometimes the cure is worse than the condition, or perhaps if it does cure one condition, it may lead to an entirely new set of undesirable conditions. That is one of the real dangers when we deal with problems that frequently come from old and indeterminate origins, and when we propose treatments that have not been proved. Sometimes there is no choice, but many times unthinking adherence to inapplicable procedures, regulations, or specifications results in counterproduction.

Although deriving from a desire to protect what we have, authoritarian attitudes and dogmatic approaches can be detrimental. For instance, rigid interpretation of legalistic regulations would deny historical status to buildings less than 50 years old, although it is obvious that some buildings or monuments are historic before they leave the drawing board. Regulations could also impose a life sentence in the literal sense wherein the existing object must remain in an unmodified condition, much like the "living dead" hospital patient tied to a cumbersome, expensive, and restrictive life-support system.

A positive attitude should permit a variety of approaches in the solution of problems and should also admit the inevitability of change and the end results of life cycles. The application of common sense and reason to problem-solving situations should provide parameters for the application of technology and limits on the expenditure of

FIGURE 6 Internal Revenue Service Building at 12th Street and Constitution Avenue, N.W. Architect: Louis Simon (1928-36). Unanticipated uses and inadequate detailing produce damaging impacts on limestone and granite arrises.

resources. Intelligent and innovative management of historic buildings and monuments should provide opportunities to make decisions in this city of decision makers—this city, which itself was determined and designed by the conscious decisions of political leaders, technicians, and public servants, many of them seekers after truth and beauty, among other things.

FIGURE 7 National Gallery of Art, East Wing Addition, at 4th Street and Washington Mall, N.W. Architect: Ioh Ming Pei (1979). Irresistible tactile attraction of sharp arris results in staining, mortar loss, and deterioration of marble at corners of affected blocks.

CONCLUSIONS

I should like to dispel any notion that I have proposed that doing nothing is the recommended form of treatment for problems. Doing nothing is simply an admission that treatment is not preferred or not possible, for whatever reasons. But benign neglect apparently has positive possibilities, too.

In the American Institute of Architects' *Guide to the Architecture of Washington, D.C.*, the authors say, "The Logan Circle is a good example of the fact that neglect is often the handmaiden of preservation. This is an eight block area of virtually unchanged large, late-nineteenth-century Victorian and Richardsonian houses focusing on the Circle itself. Nearly all were constructed during the twenty-five-year period between 1875 and 1900 as the homes of the prominent and wealthy. By the mid-eighteen-nineties taste had shifted and the Dupont Circle area had become the fashionable district, but, almost incredibly, only three of the original houses on the Circle have been demolished in the twentieth century" (1974). So here is an instance of conservation by happy accident.

If one possible definition of conservation is the mean between the extremes of complete replacement and complete loss of historic buildings and monuments, let us hope that in Washington we can practice it by constant observation, problem identification, research, invention and adaptation, planning and design, and inspection and evaluation. This conference, in its scope and its subjects, constitutes a commendable exercise toward making that practice perfect.

A GUIDE to some illustrative STONE CONSERVATION PROBLEMS AND TREATMENTS in and near WASHINGTON, DC

—— IN CONJUNCTION WITH A CONFERENCE on CONSERVATION of HISTORIC STONE BUILDINGS and MONUMENTS ——

AT THE NATIONAL ACADEMY OF SCIENCES, FEBRUARY, 1981

1. NATIONAL ACADEMY OF SCIENCES —
DETERIORATION OF MATRIX IN STONE SHAFTS OF LAMPPOSTS AT ENTRANCE

2. LINCOLN MEMORIAL —
MARBLE DISCOLORATION FROM INSECTS

3. MEMORIAL BRIDGE —
DISCOLORATION OF SCULPTURE BASES AT EAST END
DESTRUCTION OF BALUSTERS BY MOTORISTS

4. AVENUE OF HEROES —
DISCOLORATION OF MARBLE BASE AT ADMIRAL BYRD STATUE (N. SIDE)
DETERIORATION OF GRANITE WALLS AT HEMICYCLE (WEST END)

5. THE DISCUS THROWER —
STAINING AND SPLITTING OF MARBLE BASE

6. CAPITOL GATE STRUCTURES —
EXAMPLES OF EARLY USE OF "AQUIA" SANDSTONE
GATE POST IS AT 17TH STREET; GATE HOUSE IS AT 15TH STREET

7. CONSTITUTION HALL —
PROBLEM AND TREATMENT EVIDENT AT C STREET SIDE ENTRANCE

8. OLD EXECUTIVE OFFICE BUILDING
PIGEON PREVENTION DEVICES
SOME STAINING FROM COPPER FLASHINGS

9. RENWICK GALLERY —
UNSUCCESSFUL PATCHING PROJECT

10. WHITE HOUSE —
PAINT REMOVAL AT EAST END PERMITS EVALUATION OF ORIGINAL STONE

11. VETERANS ADMINISTRATION
EXTERIOR SANDBLASTED IN 1980

12. WASHINGTON MONUMENT
INTERRUPTION IN CONSTRUCTION IS EVIDENT
CLEANED TWICE IN RECENT YEARS

13. AUDITORS COMPLEX —
FORMERLY BUREAU OF PRINTING AND ENGRAVING
DETERIORATION OF BRICK & MORTAR AT REAR

14. GEN. GRANT MEMORIAL
VERY INTENSE BRONZE STAINS ON MARBLE BASES

15. CAPITOL —
DETERIORATING WEST FRONT HAS BEEN SUBJECT OF STUDIES & CONTROVERSY FOR YEARS

16. PENSION BUILDING —
SOON TO BE THE NATIONAL MUSEUM OF THE BUILDING ARTS, ITS OWN MAN-MADE MASONRY (TERRA-COTTA) HAS PROBLEMS

17. OLD PATENT OFFICE —
NOW THE MUSEUM OF AMERICAN ART, EAST STEPS HAD TO BE REBUILT BECAUSE OF SUBWAY CONSTRUCTION

18. TARIFF COMMISSION BUILDING
LIKE #17, ANOTHER ROBT. MILLS BUILDING THAT SUFFERED FROM SUBWAY CONSTRUCTION

19. APEX LIQUOR STORE —
FORMERLY THE CENTRAL NATIONAL BANK BUILDING, IT HAS EXPERIENCED SEVERE DETERIORATION OF SANDSTONE

20. FBI BUILDING —
NEW HOUSING FOR STARLINGS — OLD HOUSING CAN BE SEEN ON THE ADJACENT BLOCK OF "E" STREET AND MANY OTHER PLACES IN WASHINGTON

21. JUSTICE DEPARTMENT
CLEANED PANELS ON PENNSYLVANIA AVENUE FACADE

22. B. FRANKLIN STATUE
RECENTLY MOVED TO OLD POST OFFICE ENTRANCE FROM 10TH ST & PA. AVE, NW
LOSING FINGERS, ETC.

23. OLD POST OFFICE —
EXTERIOR CLEANED IN 1976

24. WILLARD HOTEL —
SLATED FOR RENOVATION
PAINT BEING REMOVED FROM BRICK

25. ALMAS TEMPLE —
SOME SMALL COLUMNS ON THE FACADE SHOW COMPRESSION FAILURE

26. #6 LOGAN CIRCLE
SEVERE DETERIORATION OF SERPENTINE STONE

ALSO OF INTEREST:

• NATIONAL BUREAU OF STANDARDS MASONRY TEST WALL
AT GAITHERSBURG, MARYLAND
— 30 YRS OF EXPOSURE

• CABIN JOHN VIADUCT
ON MARYLAND SIDE, NORTH OF THE DISTRICT
— STONE BELT COURSE REMOVED BY JACKHAMMER AS SAFETY MEASURE

• FORT WASHINGTON
ON MARYLAND SIDE, SOUTH OF THE DISTRICT
— RUBBLE WALLS TREATED WITH HERBICIDES

NOTE: CREDIT FOR CONTENT IS SHARED BY PAUL GOELDNER, DAVID HART, HUGH C. MILLER, ANDREA MONES, LEE NELSON & NICHOLAS PAPPAS. COORDINATION IS THE CONTRIBUTION OF WALTER ROTH.

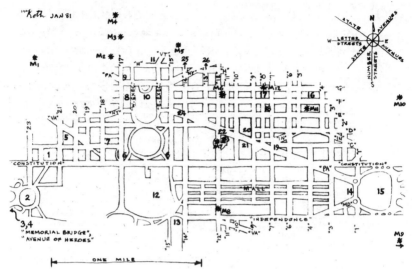

Koth JAN 81

ONE MILE

1. NAT. ACADEMY OF SCIENCES
   CONSTITUTION AVENUE
   BETWEEN 21ST & 22ND, NW

2. LINCOLN MEMORIAL
   THE MALL, WEST END

3. MEMORIAL BRIDGE
   POTOMAC RIVER,
   BEHIND LINCOLN MEMORIAL

4. AVENUE OF HEROES
   VIRGINIA SIDE, EXTENSION
   OF MEMORIAL BRIDGE

5. THE DISCUS THROWER
   STATUE IN KELLY PARK,
   21ST & VIRGINIA AVE, NW

6. CAPITOL GATE STRUCTURES
   CONSTITUTION AVE @ 17TH & 15TH, NW

7. CONSTITUTION HALL
   18TH & C STREETS, NW

8. OLD EXECUTIVE OFFICES
   17TH ST, NW - PA. TO VA. AVENUES

9. RENWICK GALLERY
   17TH ST & PA. AVE, NW

10. WHITE HOUSE
    16TH ST & PA. AVE, NW

11. VETERANS ADMIN. BLDG.
    VERMONT AVE, H TO I STS, NW

12. WASHINGTON MONUMENT
    THE MALL @ 16TH STREET

13. AUDITORS COMPLEX
    (OLD BUR. OF PRNTG. & ENGRVG.)
    14TH & C STREETS, SW

14. GEN. GRANT MEMORIAL
    THE MALL, EAST END

15. CAPITOL
    THE MALL, EAST END

16. PENSION BUILDING
    4TH, 5TH, F & G STREETS, NW

17. OLD PATENT OFFICE
    7TH, 9TH, F & G STREETS, NW

18. TARIFF COMM'N. BLDG.
    7TH, 8TH, E & F STREETS, NW

19. APEX LIQUOR STORE
    (OLD CENTRAL NAT. BANK BLDG.)
    7TH ST & PA. AVE., NW

20. FBI BUILDING
    9TH, 10TH, E STS & PA. AVE., NW
    & OLDER BLDGS NEARBY

21. JUSTICE DEPT. BLDG.
    9TH, 10TH STS, ALONG PA. AVE., NW

22. B. FRANKLIN STATUE
    12TH ST & PA. AVE., NW

23. OLD POST OFFICE
    12TH ST & PA. AVE, NW

24. WILLARD HOTEL
    14TH ST & PA. AVE., NW

25. ALMAS TEMPLE
    K ST, BETWEEN 13TH & 14TH, NW

26. #6 LOGAN CIRCLE

METRO STATIONS:

M1 = FOGGY BOTTOM
M2 = FARRAGUT WEST
M3 = FARRAGUT NORTH
M4 = DUPONT CIRCLE
M5 = McPHERSON SQUARE
M6 = METRO CENTER
M7 = FEDERAL TRIANGLE
M8 = SMITHSONIAN
M9 = CAPITOL SOUTH
M10 = UNION STATION
M11 = JUDICIARY SQUARE
M12 = GALLERY PLACE

Geological Sources
of Building Stone

NORMAN HERZ

Throughout history, the geologic availability of materials has been a principal factor affecting techniques of construction, structure and stability, decorative detail, and overall aesthetic aspects of buildings and monuments. Greece and Rome, for instance, had abundant, locally available stones, including marble and limestone, all of which were used in the construction of the principal monuments of classical times. In the United States, weathering-resistant, strong, and attractive building stones are abundant in all but one geological province: the Atlantic and Gulf Coastal Plain. Abundant stone is supplied by the other geological provinces, which are composed of crystalline rocks or older sediments. Four states accounted for 51 percent of the total U.S. production of dimension stone in 1980: Indiana, a leader in limestone, and Georgia, Vermont, and New Hampshire in marble and granite. Granite accounted for 50 percent of domestic production; the rest, in order of volume, were limestone, sandstone, slate, and marble.

Throughout recorded history man has had a special relationship with building stone. He was quite content to construct his own home out of timber, mud wattle, thatch, or whatever other materials were easily worked and available; but for his monumental buildings, only the most beautiful and durable stone would do. We think of ancient Athens as a city of gleaming marble, of the monumental buildings of the Acrop-

Norman Herz *is Professor of Geology, Department of Geology, University of Georgia, Athens.*

olis and of the Agora, when in all probability its principal aspect must have been that of unbaked, sun-dried brick—the adobe of our southwestern United States—capped by terra-cotta roof tiles. Wycherley has speculated that the view from the Acropolis today, of red-tile roofs, must be identical to that of classical times.[1] What has lasted over two millenia are not the adobe dwellings but the buildings and monuments constructed of gleaming white marble from Mount Pentelikon in Attika or from the island of Paros in the Cyclades. The ancient Greeks sought to preserve and transmit to later generations what they deemed most significant in their culture—not their mean day-to-day existence but their philosophy, religion, and form of government; and this could best be done through marble monuments.

Almost all important cultures in both the Old World and the New World have shared this special feeling about building stone. Primitive societies set up stone cairns to memorialize the site of important events or to mark important routes. More advanced cultures quarried and used dimension stones to construct monuments and buildings, again with the thought of preserving what they thought most important in their heritage. In this century we have moved steadily from buildings constructed principally of dimension stone to those using materials that are easier and cheaper to handle, such as steel and concrete. However, when we deem a building monumental in scope, such as the new National Gallery of Art, marble is still used. In many architectural designs, dimension stone is used as a thin curtain wall or veneer, in slabs 7/8 in. to 5 in. thick, to add grace and beauty to what may appear otherwise aesthetically less attractive constructions.

The term "stone" as used in this country includes all consolidated rock that is mined or quarried and used for construction, roads, or chemical, metallurgical, and agricultural activities. As a construction material, dimension stone is prepared to predetermined size and finish. About 314 B.C., Theophrastus listed characteristics that made building stone valuable: (a) found in large areas and made up of whole layers; (b) can be extracted in whole blocks; (c) possesses a pleasing color and other aesthetic features, such as (d) smoothness; and (e) is relatively hard, i.e., structurally strong.[2] In the 2300 years since these observations, all we have added is the ability to resist the accelerated weathering common in today's urban or acid-rain environment.

The availability of building stones was an important architectural consideration until the start of this century. Abundant limestone and marble in classical Greece enabled the architect to develop a unique style that combined elaborate use of marble with delicate decorative detail in all visible parts of the buildings. Limestone and marl blocks were used to lighten the building superstructure, but they were always veneered with marble. The Romans, on the other hand, developed massive structures that generally employed less marble and more concrete, although all building materials of the known world were available to and used by Roman architects. Rome developed many concepts and systems, such as mass production, accumulation of stocks, prefabrication, and standardization of qualities and dimensions of building stones and other materials. This made it possible to accomplish such feats as building the Pantheon in less than 10 years (ca. 118–128 A.D.) and the Baths of Diocletian in less than 8 (ca. 298–306 A.D.).[3]

The flowering of the Gothic cathedral was related directly to the availability of attractive and easily quarried sedimentary rocks. In the Middle Ages the cost of quarrying and finishing the stone equalled the cost of transporting it 12 miles from the quarry site, so that local stone generally dictated the architectural style,[4] but important exceptions always existed for the most prestigious buildings. Much of the stonework for the Norman cathedral at Canterbury, for example, was obtained from quarries in Caen, France. As our own capital has grown in prestige and international importance, we have also used a greater variety of exotic building stones, such as Carrara marble and travertine from Italy, larvikite from Norway, and anorthosite from Canada.

VARIETIES OF BUILDING STONES

Almost every kind of rock can be used as dimension stone.[5] The principal controls on usage are aesthetic appeal and physical properties, including resistance to weathering. The geological definition of a rock is based on its chemistry, fabric, and mineralogy; these attributes are also the principal determinants of its properties. The American Society for Testing and Materials (ASTM) has adopted standard definitions for the principal commercial dimension stones. Building stones are described below both from the geological and ASTM points of view.

Rocks are divided into three overlapping genetic groups:

- Sedimentary rocks, such as limestone and sandstone.
- Igneous rocks, such as granite and diabase (traprock).
- Metamorphic rocks, such as marble and slate.

Limestone

ASTM defines limestone as a rock of sedimentary origin, composed principally of calcium carbonate (calcite) or the double carbonate of calcium and magnesium (dolomite). The textures vary greatly, from uniform grain size and color to a cemented-shell mash. Oolitic limestone, a popular building stone in this country, Britain, and France, consists of cemented rounded grains of calcite or aragonite generally under 2 mm in diameter. Some limestones have varying amounts of other material, such as quartz sand or clay mixed in with the carbonate minerals. Most limestones are formed of shells or reworked shell fragments, although many commercial limestones, including oolitic and very fine-grained and compact varieties, are chemical precipitates.

Sandstone

ASTM defines sandstone as "a consolidated sand in which the grains are composed chiefly of quartz and feldspar, of fragmental texture, and with various interstitial cementing materials, including silica, iron oxides, calcite, or clay." Commercially used sandstone is a clastic sediment consisting almost entirely of quartz grains, 1/16 to 2 mm in diameter, with various types of cementing material. Enough voids generally remain in the rock to give it considerable permeability and porosity. In the United States, commercially available sandstones include the well-known brownstone, an arkosic sandstone that is rich in feldspar grains and was quarried in the Triassic basins of the eastern states.

Travertine

Travertine is a variety of limestone deposited from solution in groundwaters and surface waters. When it occurs hard and compact and in extensive beds, as around Rome, it can be quarried and used as an attractive building stone. It is generally variegated gray and white or buff, with irregularly shaped pores distributed throughout the groundmass.

Granite

Commercial granite includes almost all rocks of igneous origin. True granites consist of alkali feldspars and quartz with varying amounts of other minerals, such as micas and hornblende, in an interlocking and granular texture, and with all mineral constituents visible to the naked eye. Geologically, granite is distinguished from other rocks that it resembles, such as granodiorite, quartz monozonite, and syenite, on the basis of the percentages of quartz, potassium feldspar, and plagioclase feldspar. This distinction is not made commercially; in fact, black fine-grained igneous rocks, such as basalt or diabase, are commonly called "black granite." Other dark "granites" include rocks that, petrographically, are anorthosite, gabbro, syenite, and charnockite.

Marble

According to ASTM, commercial marble includes all crystalline rocks composed predominantly of calcite, dolomite, or serpentine and capable of taking a high polish. Geologically, marble is considered only as a metamorphic rock formed by the recrystallization of a limestone or dolomite under relatively high heat and pressure. Thus, in addition to geological marble, commercial marble includes many crystalline limestones, travertine, and serpentine, a metamorphosed ultramafic rock. In the metamorphic process, original sedimentary features, except for bedding, which is preserved as a compositional layering, are destroyed. The original minerals, calcite and dolomite, are recrystallized in an interlocking mosaic texture, and the impurities form magnesium and iron silicates. The color of many marbles is due to these accessory minerals, such as talc, chlorite, amphiboles, and pyroxenes, as well as iron oxides, hydroxides, sulfides, and graphite.

Slate

ASTM requires a slate to possess an excellent parallel cleavage that allows the rock to be split with relative ease into thin slabs. Slate is a metamorphosed rock derived from argillaceous sediments consisting of extremely fine-grained quartz, the dominant mineral, and mica and other platy minerals. The color of slate is generally determined by the oxidation state of the iron or the presence of graphite or pyrite.

Other Types

A great variety of other types of rocks are sold commercially, including: (a) quartzite—a metamorphosed sandstone consisting almost entirely of quartz and utilized locally, as the Sioux Falls quartzite of South Dakota and the Baraboo quartzite of Wisconsin; (b) greenstone—defined by ASTM as a metamorphic rock principally containing chlorite, epidote, or actinolite; (c) basalt or traprock—a microcrystalline volcanic or dike rock that consists primarily of pyroxene and a calcic plagioclase (the stark black churches of the Auvergne of Central France are largely made of basalt); and (d) obsidian—a volcanic glass that commercially includes pumice in the United States, has low density because of its frothy texture, and can be easily shaped with hand tools.

BUILDING STONE IN THE UNITED STATES

In the early part of this century the total value of dimension stone produced was greater than that of crushed stone.[6] As other materials were substituted for dimension stone in construction, and as a national highway program was developed, these roles were reversed. In recent years the total amount of dimension stone has been less than 0.5 percent of total stone produced and its value about 4 percent of the total. Despite this decline in relative production, the actual amount of dimension stone produced since 1973 has not varied greatly from 1.5 million tons. A high of 1.9 million tons was reached in 1974 and a low of 1.3 million tons in 1980. The total value of dimension stone increased from $86.0 million in 1973 to $147 million in 1981. (See Table 1.)[6,7]

The total production of dimension stone in 1981 increased less than 1 percent over 1980 but its value rose 6 percent. In 1981 in terms of volume, about 50 percent of the production was granite, 22 percent limestone, 13 percent sandstone, 5 percent marble, and 7 percent slate. In 1979 marble was the most costly at $177 per ton, followed by slate, $147; granite, $110; limestone, $55; and sandstone, $40. In 1980, dimension stone was produced in 39 states by 263 companies operating 426 quarries. Leading states were Georgia, which produced 18.1 percent of U.S. building stone; Indiana, 13.4 percent; Vermont, 13.4 percent; and New Hampshire, 6.3 percent. These four states accounted for just

TABLE 1   U.S. Production of Dimension Stone, 1929–1981

Year	Tons Produced (millions)	Value (millions of dollars)
1929	4.7	70
1939	2.3	25
1949	1.8	52
1959	2.3	25
1969	1.9	99
1979	1.35	123
1981	1.32	147

SOURCE: U.S. Bureau of Mines, 1975, 1982.

over half of domestic production. Other states producing more than 35,000 tons in 1980 include California, Massachusetts, Minnesota, North Carolina, Ohio, Pennsylvania, South Dakota, Texas, and Wisconsin.

Domestic reserves of building stone must be considered inexhaustible,[5] although shortages of special varieties may develop because of economic or aesthetic factors. Geological controls of various types of rocks are such that all common building stones occur abundantly in nature, so that the exhaustion of one quarry or district can quickly lead to the discovery and exploitation of similar stone in another district. The factors that determine whether a new building stone district should be opened are economic and environmental: Can the stone be quarried, finished, and transported to markets under existing environmental regulations and at competitive prices? The most important markets have been the more densely populated areas—the Northeast, upper Midwest, and Southern California—but this picture is changing as people move to the Southwest, or Sun Belt. The list of the principal producing states shows that the dimension stone industry has been concentrated in areas where the geology is favorable, but also where traditional markets are not too far removed.

The distributional and geological controls over building stone production in the United States are best understood from the point of view of lithologic provinces (Figure 1). Five provinces can be distinguished, corresponding roughly to physiographic provinces, that control the lithologic types found in each.[5] These are (a) Atlantic and Gulf Coastal Plain, (b) Appalachian Crystalline Province, (c) Central-Interior Sedimentary Basins, (d) Lake Superior Crystalline Province, and (e) Western Province. The Appalachian and the Central-Interior provinces have produced the greatest amounts of building stones, being blessed with both favorable geology and proximity to markets.

Atlantic and Gulf Coastal Plain

The Atlantic and Gulf Coastal Plain Province commences in the southeastern half of New Jersey and includes part or all of the coastal states to Texas. It is generally underlain by poorly consolidated sedimentary rocks of Cretaceous to Recent age that lie in nearly horizontal layers. Very little dimension stone has been produced because the rocks lack strength, although in Colonial times a coquina limestone that hardened when exposed to the air was widely used in Florida and other southeastern states.

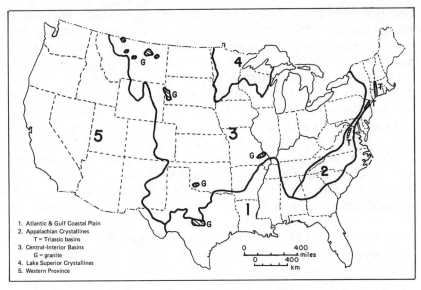

1. Atlantic & Gulf Coastal Plain
2. Appalachian Crystallines
   T = Triassic basins
3. Central-Interior Basins
   G = granite
4. Lake Superior Crystallines
5. Western Province

FIGURE 1 Lithologic provinces of the United States. SOURCE: Laurence, 1973.[8]

Appalachian Crystalline Province

The Appalachian Crystalline Province includes the Appalachian mountain belt, from the Blue Ridge and Piedmont in the south, northward to the Reading Prong and New Jersey–Hudson Highlands, the Adirondack Mountains, and New England. Included geographically within this province is the lithologically unrelated Triassic Basin Subprovince.

This province consists of crystalline rocks, both igneous and metamorphic, of Precambrian and Paleozoic age that formed under relatively high pressures (i.e., deep burial) and high temperatures. Uplift and erosion have exposed this ancient "root zone" of what must have been at one time much higher mountain ranges. This province leads in the production of granite, slate, marble, serpentine, and other crystalline rocks and has the added attribute of proximity to important markets. Slate is principally produced in Virginia, Pennsylvania, New York, and Vermont; marble in Georgia, Vermont, and Alabama; and serpentine and verde antique in Vermont and Virginia. Granite has been produced in almost every state within the province, principally in Georgia but also in Connecticut, Rhode Island, Massachusetts, New

Hampshire, Vermont, New York, Pennsylvania, North Carolina, and South Carolina.

The Triassic Basin Subprovince was formed before the continental breakup of North America and Europe–Africa. It consists of down-faulted basins with tilted but undeformed sedimentary rocks, as well as basaltic flows and diabasic dike rocks (traprock). The three principal basins are (a) along the Connecticut River, where a large production of brownstone, an arkosic red sandstone, was used in building construction in many large eastern cities around the turn of the century (production now is negligible because the stone does not resist the accelerated weathering of modern city environments); (b) the Newark Basin of New Jersey, which has produced much traprock; and (c) the Triassic Basins of Virginia.

Central-Interior Sedimentary Basins

The Central-Interior Sedimentary Basins are generally underlain by flat-lying sedimentary rocks of Paleozoic, Mesozoic, and Cenozoic age and also include a few very important areas of crystalline rocks. The province extends from the western slopes of the Appalachian Mountains to the foothills of the Rocky Mountains. The only areas where the rocks are not flat-lying, but are rather complexly folded and faulted, are in the Valley and Ridge Physiographic Province, just west of the Blue Ridge, and the Ouachita Mountains of Oklahoma and Arkansas. Crystalline rocks are exposed in the cores of domal uplifts, as in the Black Hills of South Dakota and elsewhere.

The eastern two-thirds of the province is underlain by Paleozoic sedimentary rocks that have produced more than 80 percent of the total domestic sandstone and limestone. Good geological controls are evident in the distribution of the quarried stones. The province was a vast inland sea during most of the Paleozoic and bordered on the ancient Appalachia landmass. Nearshore, detrital sandstones were laid down; to the west, in the shallow seas away from the ancient shoreline, carbonate limestones were precipitated. Thus sandstone is produced in New York, Ohio, and Pennsylvania, closer to the ancient shoreline, and the largest limestone production is farther west, in Indiana and Missouri.

Granite exposed in this province is an important source of building stone and is produced in South Dakota, Missouri, Oklahoma, and Texas.

Lake Superior Crystalline Province

The Lake Superior Crystalline Province comprises parts of Minnesota, Wisconsin, and Michigan at the southward terminus of the much greater Canadian Shield Province of Precambrian age. This small U.S. province has produced much granite; in fact, it stands second after the Appalachian Crystallines in production.

Western Province

The Western Province includes the area from the Rocky Mountain foothills to the Pacific Ocean. Geologically, there are a great number of lithologic subprovinces represented that formed in many different environments. A great variety of rocks can be found; the sole limitation on dimension-stone production from this area is the distance to viable market centers.

All the principal types of building stones are produced in this area. Granite is the most widespread and has been quarried in California, Colorado, and Washington. Sandstone has come principally from the Colorado Plateau and limestone from California. Travertine has been quarried in Idaho and obsidian and pumice in California and Nevada. Much of the production has been near the large population centers of southern California.

FOREIGN SOURCES

Tariffs on dimension stone vary from zero to 12.7 percent ad valorem for most-favored nations, according to type, size, value, and degree of preparation. In the period 1978–80 in terms of value, Italy supplied 71 percent of imports of dimension stone, Canada 9 percent, Mexico 5 percent, and all other countries 15 percent. In terms of total value, we export less than half of what we import: In 1980, exports were $36.4 million and imports $88.9 million. Italy dominates the international trade in dimension stone; it is the leading importer of rough blocks and the leading exporter of finished stone. The principal export has been marble, especially from the Carrara district; but travertine from around Tivoli, serpentine, volcanic tuff, and limestone are also exported. Other countries have supplied many types of granite, true granite as well as the so-called black granite of Canada (anorthosite from Lac Saint Jean) and of Norway (larvikite).

TECHNOLOGY

The dimension stone industry is labor intensive, has been so in the past, and must remain so in the foreseeable future. This has hampered growth and more widespread use of building stone, since production methods remain slow, cumbersome, and costly.[9] Research on extractive and finishing equipment has recently resulted in improvement of basic designs and has helped keep the cost of building stone lower than expected in terms of normal inflation. Modernization has involved the use of high-speed, diamond stone saws, grinders, and drills, as well as improved conventional wire saws and jet or water-piercing drills. Little blasting can be done in mining dimension stone, and quarrying must involve the use of diamond saws, wire saws, and drilling machines. Large circular saws are used for final processing; some are 10 ft (3 m) or more in diameter with diamond or steel shot inserts. Similar equipment, utilizing diamond saws and abrasives, is employed for the final polishing and decorating of the stone.

ENVIRONMENTAL PROBLEMS

Environmental problems confront stone producers to a greater extent than they do producers of almost any other mineral commodity.[6] The problems arise not from noxious waste products but because many quarries are in urban areas, close to their potential markets. Such problems include high noise levels from quarrying operations, dust from mining and sawing, solid waste blocks left on the surface, and general unsightliness. Rarely, as happened in a serpentine quarry in Maryland, particulate asbestos derived from asbestos fibers in the rock is detected in air. The question of possible health hazards from such sources has been studied by both the EPA and the Bureau of Mines, which has established a Particulate Mineralogy Unit to examine further and classify rock-derived dusts and fibers. None of these environmental problems is insoluble, except possibly those relating to serpentine, which is one of the least important of the building stones.

The single major threat to stone resources is probably competing land uses. Once the deposits have been covered by building lots or made into parks or recreational areas, they are removed as potential sources of raw materials. Large eastern cities such as New York, which must now obtain much of its sand and gravel for construction from the floor of the continental shelf, are prime examples of the consequences of unlimited development without planning for future raw material sources. Land rehabilitation after depletion of dimensional stone deposits by surface mining can also be very costly. In the average mining operation, ore is extracted, and the waste, generally most of the mined rock, can be returned to the hole left by quarrying to help recontour the surface. In a dimension stone quarry much of the stone removed is used, and restoration of the land to its previous surface using on-site wastes is all but impossible.

OUTLOOK

The demand for dimension stone has been relatively stable or decreasing slightly since 1962. The only way that the competitive position of stone relative to other construction materials could be improved would be to develop low-cost methods of extraction and maintain consistent standards of color and strength in large-scale, rapid construction projects. In many parts of Georgia, for instance, granite is a much stronger and cheaper curbstone material than concrete because of improved production methods.

Stone may develop discoloration or weathering soon after being put in place in a monument or construction.[10] Methods for more rapid testing of dimension stones are needed. Such tests must also include exposure of the stone to various types of environments.

Conservation of building stone, in the sense of conservation of a natural resource, is not a problem. Our domestic resources are inexhaustible, although real or potential shortages in localized areas can be serious. Once an area has been identified as the source of a strong and attractive building stone, every effort should be made to preserve it by zoning laws and by conscious efforts to discourage urban development within it. Geological studies by the federal and state governments should always identify potential sources of building stones. Perhaps a data base can be built up of future sources, especially deposits that merit special legal protection so that they can be preserved until needed.

So long as man believes in preserving his cultural heritage and transmitting his monuments to future generations, his primary concerns must remain the preservation, production, and protection of durable and attractive building stones.

REFERENCES

1. R.E. Wycherley, *The Stones of Athens* (Princeton, N.J., 1978).
2. A. Dworakowska, Quarries in Ancient Greece, Polish Academy of Sciences, *Bibliotheca Antiqua*, 14 (1975).
3. J.B. Ward-Perkins, *Roman Architecture* (H.N. Abrams: New York, 1977).

4. P. McCleary, Structure and Intuition, *American Institute Architect Journal*, 69(12), pp. 56–59, 119 (1980).

5. W.R. Power, Dimension and Cut Stone, pp. 157–174 in *Industrial Minerals and Rocks*, S.J. Lefond, ed. (American Institute of Mining Engineers: New York, 1975).

6. J.E. Shelton and H.J. Drake, Stone, pp. 1031–1048 in *Mineral Facts and Problems, 1975* (U.S. Bureau of Mines: Washington, D.C., 1976).

7. U.S. Bureau of Mines, *Stone in 1981, Mineral Industry Surveys* (U.S. Bureau of Mines: Washington, D.C., 1982).

8. R.A. Laurence, Construction Stone, in *U.S. Geological Survey Professional Paper 820*, pp. 157–162 (1973).

9. A.H. Reed, *Stone: Mineral Commodity Profile 17* (U.S. Bureau of Mines: Washington, D.C., 1978).

10. E.M. Winkler, *Stone: Properties, Durability in Man's Environment*, 2nd edition (Springer-Verlag: New York, 1975).

Physical Properties
of Building Stone

EUGENE C. ROBERTSON

Porosity and permeability seem to be the most important physical properties affecting weathering and deterioration of building stones by water and gases. Thermal and mechanical properties, because of their effects on permeability, chemical reactions, strength, and stability, can be important in diagnosing decay processes in stones; but optical, electrical, and magnetic properties have little significance in deterioration processes. Measurement of physical properties on small laboratory samples of any rock type range widely because the composition and character of every rock type differ according to locality. Variations in physical properties because of compositional and textural inhomogeneities can be seen in small to large quarried blocks or in rock in place and can be more significant in explaining rock deterioration than laboratory tests of the physical properties of selected small samples of rock. Examples of inhomogeneities are intercalated shaley layers, calcite, limonite, or clay cements; thin to thick bedding; mineral variations within beds in sedimentary rocks; foliation; induration; microfracturing; and incipient to open jointing.

Certain physical properties of building stones are very important in determining the susceptibility of stones to natural weathering or deterioration caused by pollution, whereas other properties have negligible influence. Most physical properties are discussed in this paper,

Eugene C. Robertson *is Geophysicist, U.S. Geological Survey, Reston, Virginia.*

and distinctions are drawn between physical properties measured on small laboratory samples and those observed on stone in place or in large blocks. Chemical properties of building stones need to be considered in conjunction with physical properties in studying processes of deterioration, and they are covered elsewhere in these proceedings.

The comprehensive book on properties and durability of stone by Erhard Winkler[1] has been helpful in preparing this report. Other books containing data on physical properties of building stones are those of Bowles,[2] Merrill,[3] Schaffer,[4] and Winkler.[5]

PHYSICAL PROPERTIES

Samples of one type of rock obtained from different localities, inevitably differ considerably in their properties because of variability in the composition and texture of the rock among the localities. Even though the chosen samples with the same rock name might appear uniform, their measured properties would vary so widely that an average value for that rock type would be misleading. Therefore, it seems appropriate to give only ranges of values for physical properties of common building and monument stones (Table 1). The names of rocks are quite general in their geologic usage (as in this report), and are even less specific in stone industry usage. For a few rocks for which only a few measurements have been published (e.g., soapstone and serpentinite), the range limits in Table 1 were estimated by comparing values with those of other types of rocks.

Porosity and permeability are probably the most important physical properties of rocks for studies of decay and corrosion of building and monument stones. This is because these properties characterize the accessibility of water to the interior of the stones and because water in all of its three phases is perhaps the most important substance affecting the weathering and deterioration of the stones. Thermal and mechanical properties are next in importance in the decay of rocks. Optical, electrical, and magnetic properties have very little importance.

Physical properties depend primarily on the origin and geologic history of each rock. Because mineral composition and texture differ according to the varying geologic histories of rocks, the values of physical properties range widely (Table 1). Building stones are polycrystalline mineral aggregates, not single crystals. Thus intergranular bonding, pore shape and size, and fabric are more important than the physical properties of individual mineral grains, even including their anisotropy.

TABLE 1 Ranges of Values of Physical Properties^a of Building Stones ^{1,7,8,9,10,11}

Rock Type	Aggregation Properties			Thermal Properties			Mechanical Properties			
	ρ (g/cm³)	ϕ (%)	μ^b (−log d)	α (1/10⁶ °C)	K (mcal/cm s °C)	k (10⁻² cm²/s)	H	E (Mb)	S (kb)	R (kb)
Granite	2.5–2.7	0.1–4	9–6	5–11	3–10	0.5–3	5–7	0.3–0.6	0.8–3.3	0.1–0.7
Gabbro	2.8–3.1	0.3–3	7–5	4–7	4–6	1–2	5–6.5	0.5–1.1	1.1–3.0	0.1–0.7
Rhyolite–Andesite	2.2–2.5	4–15	8–2	5–9	2–9	0.4–3	5–6.5	0.6–0.7	0.6–2.2	0.01–0.7
Basalt	2.7–3.1	0.1–5	5–1	4–6	2–5	0.4–1.5	4–6.5	0.5–1.0	0.5–2.9	0.1–0.9
Quartzite	2.5–2.7	0.3–3	7–4	10–12	8–16	2–8	4–7	0.6–1.0	1.1–3.6	0.1–1.0
Marble	2.4–2.8	0.4–5	6–3	5–9	3–7	0.5–1.5	2–4	0.2–0.7	0.4–1.9	0.04–0.3
Slate	2.6–2.9	0.1–5	11–8	8–10	3–9	0.5–3	3–5	0.3–0.9	0.5–3.1	0.05–1.0
Sandstone	2.0–2.6	1–30	3–0	8–12	2–12	0.4–5	2–7	0.03–0.8	0.2–2.5	0.01–0.4
Limestone	1.8–2.7	0.3–30	9–2	4–12	2–6	0.4–1.5	2–3	0.1–0.7	0.2–2.4	0.1–0.5
Shale	2.0–2.5	2–30	9–5	9–15	1–8	0.3–2	2–3	0.1–0.4	0.3–1.3	0.02–0.5
Soapstone	2.5–2.8	0.5–5	6–4	8–12	2–7	0.4–1.5	1	0.01–0.1	0.1–0.4	0.01–0.1
Travertine	2.0–2.7	0.5–5	5–2	6–10	2–5	0.4–1	2–3	0.1–0.6	0.1–1.5	0.02–0.1
Serpentinite^c	2.2–2.7	1–15	7–3	5–12	3–9	0.5–3	2–5	0.1–0.5	0.7–1.9	0.05–0.1

^{a} The symbols used as column headings are: ρ, bulk density in d; α, linear thermal expansion of millionths of reciprocal degrees Celsius; K, thermal conductivity in millicalories per centimeter-second-degree Celsius; k, diffusivity in hundredths of square centimeters per second; H, hardness by Mohs' scale; E, Young's modulus of elasticity in megabars; S, uniaxial compressive strength in kilobars; and R, modulus of rupture in kilobars.

^{b} Permeability units are negative logarithms; thus 9 means 10^{-9} darcies.

^{c} Serpentinite is much like verde antique stone.

Aggregation Properties

The density, porosity, and permeability of rocks are different measures of the state of aggregation of the mineral grains that make up the rock and are measures of the accessibility of fluids, principally air and water, into and through the rock. Geologic history determines the aggregation.

Density

Grain density, ρ_G, is the ratio of grain mass to grain volume of rock having no porosity. The values in Table 1 are bulk densities, ρ_B, which are the mass of grains divided by pore volume plus grain volume. Measurements of ρ_G by immersion methods on whole samples can be in error by as much as 10 percent owing to incomplete saturation of inaccessible pores; ρ_B measurements would be more accurate. The ρ_G measured on a crushed sample would give a more reliable value.

Moen put lower and upper density limits of 1.7 to 2.2 g/cm³ on commercial stone that would be favorable for preparation and working.[6] By his criteria, stones having a density greater than 2.2 g/cm³ are too hard to work easily with masonry tools, and stones having a density less than 1.7 g/cm³ are too soft and easily weathered. However, those having densities above 2.2 g/cm³ resist weathering better than stones with lower densities and can be worked with modern abrasives and machines.

Porosity

Porosity, ϕ, is the ratio of pore volume to bulk volume. Porosities of common building stones are listed in Table 1. Porosity of igneous and metamorphic rocks is low, usually less than 5 percent, but it can be as high as 40 percent for sedimentary rocks. Pores are important in rock decay because they are receptacles for fluids and sources of weakness for ambient stresses. Hudec found that weathering of stone is enhanced by decrease in pore size.[5]

Mineralogy and degree of metamorphism cause sizes and shapes of pores to differ in sandstone and shale and to differ in quartzite and slate. Clay and mica minerals make up half of most samples of shale and slate; they occur as closely packed, parallel flakes, which make for tabular and very small pore spaces, about 0.01 μm across in most shales. Porosities of shale and slate range from 30 to 0.1 percent, depending on the degree of compaction, diagenesis, and metamorphism. In sandstones, porosity does not depend on grain size but does depend on sorting of the original sand grains. For example, in unconsolidated sand beds, porosity ranges from 40 percent in very well sorted sands to 25 percent in very poorly sorted sands. In addition, compaction of sandstone and deposition of silica or other mineral cement in quartzite can reduce markedly the pore size, porosity, and permeability (Table 1).

Permeability

The ease of flow of fluid through a rock is defined empirically by Darcy's law, in which the flow depends on the permeability of the rock and on the pressure and viscosity of the fluid:

$$Q = \mu P/v \ (L/A), \tag{1}$$

where Q is discharge in cubic centimeters per second, μ is permeability in darcies, P is pressure difference in bars, v is fluid viscosity in centipoises, L is distance of flow in centimeters, and A is cross-sectional area in square centimeters.

Intrinsic permeabilities obtained by laboratory measurements of intact samples of building and monument stones are listed in Table 1. Joints and fractures increase the permeability of rocks in place at shallow depths by 10 to 100 times the intrinsic values for samples of solid rock given in Table 1. This increase is described in a very comprehensive review of the permeability of rocks.[12]

Water at 200° C, flowing through granite having 10^{-3} darcy crack permeability, dissolved and reprecipitated enough silica in one month to reduce the permeability to 10^{-4} darcy.[13] Silicate and carbonate rocks are susceptible to such sealing by dissolution and reprecipitation of silica and carbonate ions.

Most research on intrinsic permeability has been performed on sedimentary rocks; workers have studied the effects of porosity, pore size and shape, and mineralogy. Data were obtained in studies of the migration of oil and gas. Very few equivalent data are available for igneous and metamorphic rocks. The permeabilities of stones range from several hundred darcies in river sand, through 0.1 darcy for common sandstone of 20 percent porosity, to 10^{-9} darcy for common shale. Clay minerals in shale or sandstone reduce permeability markedly because the pores between clay particles are very small; smectite clays in a rock expand by water absorption, further reducing permeability.

Figure 1 presents points from measurements on five sandstones in rows labeled by their porosities, which range from 8 to 22 percent;

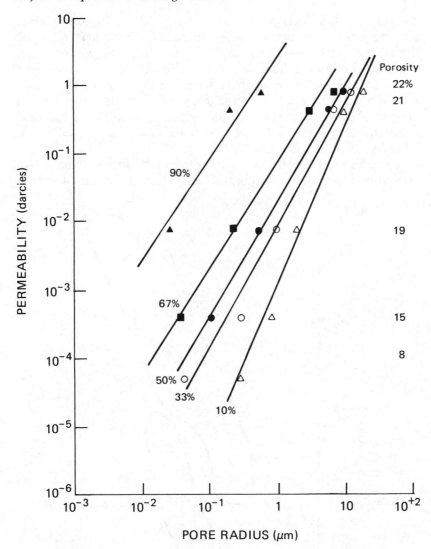

FIGURE 1 Effect of pore size on permeability of five sandstones having porosities as shown; a mercury injection technique was used on cores. The symbols aligned horizontally with each porosity percentage in the right column are for measurements on a separate rock. The percentages marked on the lines represent the proportion of pores larger than the pore radius for a given permeability on the ordinate axis (redrawn from Blatt et al.).[14]

the permeability, μ, is plotted against pore size, eliminating grain size, sorting, and cement characteristics. The percentages on the lines show relatively how many pores are larger for any point on the line. The line on the right is for sandstones in which 10 percent of the pores are larger than those found on the abscissa for a given μ on the ordinate; the line on the left is for sandstones in which 10 percent of the pores are smaller for a given μ. As might be expected intuitively, the permeability of sandstones varies exponentially with porosity and with pore size.

Figure 2 shows the effects resulting from differing contents of two clay minerals on the permeabilities of several sandstones of similar

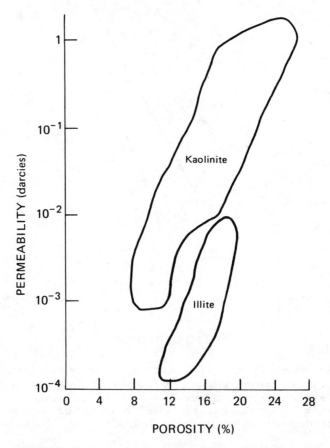

FIGURE 2 Variation of permeability with two types of clay contained in several sandstones of similar porosities (modified from Blatt et al.).[14]

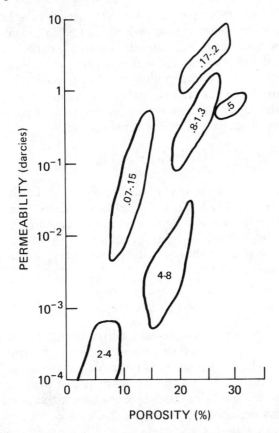

FIGURE 3   Variation of permeability with specific internal surface areas (shown in cm^2/cm^3 inside enclosed areas) of six sandstones having a range of porosity (modified from Blatt et al.).[14]

porosities. The specific surface areas (cm^2/cm^3) of six sandstones of varying porosity are shown inside the enclosed areas in Figure 3. Specific surface area is high for fine-grained, low-permeability sandstones; it is low for coarse-grained, high-permeability sandstones.

Physical Models for Aggregation Properties

The intrinsic density, porosity, and permeability of a sample would appear to be closely related, judging from their definitions. However, measured values of these properties differ from absolutely accurate values enough so that they cannot be calculated exactly from each other. The differences are probably due to the effects of isolation, small size, and irregular shape of some pores, leading to incomplete saturation by the measuring fluid and diminished accuracy of measurement. The following discussion and equations are meant to provide the reader with some understanding of what physical characteristics are important and to permit calculations for comparison purposes.

The porosity, ϕ, of a dry rock—that is, the ratio of pore volume to bulk volume—is given in terms of densities by:

$$\phi = (\rho_G - \rho_B)/\rho_G. \tag{2}$$

A good value of pore volume is needed to obtain ρ_G, but it is not easy to measure closely. However, as ρ_G can be calculated from the rock's mineral composition, and as ρ_B is more easily measured, a reasonable estimate of ϕ can be calculated from equation 2.

A relation between porosity, ϕ, and permeability, μ, was found empirically by Kozeny:[14]

$$\mu = 10^6\phi/2t^2s^2, \tag{3}$$

where μ is in darcies, ϕ is in percent, t is tortuosity (usually taken as 2 to 3), and s is specific surface area in cm^2/cm^3 of grains in a rock. The interdependence of μ and ϕ is not clearly understood. The value of s is obviously strongly affected by grain size, so that μ for a coarse-grained sandstone can be an order of magnitude higher than for a fine-grained sandstone, although both have the same ϕ.[14] The Kozeny formula has limited use because s and t are difficult to estimate.

Absorption of fluids in rocks depends on the connected, effective (i.e., permeable) porosity. Connected pores in building stones can be visualized as a system of capillary passages in which the surface tension of water becomes important. The surface tension γ of a fluid in a capillary crack of width d is given by:

$$\gamma = C\,d\,h, \tag{4}$$

where C is a constant and h is the height of rise of the fluid. For a given fluid, γ and C would be fixed; therefore, the smaller the crack width d the greater the rise h in a capillary passage. Crack widths of 5 μm have been measured in building stones; one epoxy that was injected into the pores of deteriorated stones to try to seal them has itself been found to contain crack widths of 2 to 10 μm. Lewin discusses capillary flow in detail in these proceedings. He points out that the volume of flow is proportional to the radius of the capillary to the fourth power. Saline water has been observed to rise 4 to 10 m in capillaries in sandstone and other masonry materials.[15]

THERMAL PROPERTIES

The effects of diurnal and seasonal heating and cooling on deterioration of building and monument stones can be significant on a microscopic scale. These effects involve conduction of heat by solids and induced thermal stresses. Some of the measurements that have been made of thermal stresses and their effects on rock in place are described below.

Thermal Expansion

Thermal stresses resulting from changes in the temperature of ambient air can be large enough to produce microfractures in and between the mineral grains of a rock. This can happen even in temperate climates because of anisotropy and differences in the thermal expansions of the minerals. An important feature is that the fracturing is irreversible, and thereafter the permeability will be greater and will allow greater penetration of water.

Ide found that although the volume of several common rocks did not increase perceptibly upon heating, microfractures formed by differential expansion of mineral grains, resulting in a very marked and irreversible decrease in the elastic modulus.[16] For example, a 25-fold reduction, from 0.8 to 0.03 M bar, in the elastic modulus, E, of a granite resulted from heating to 500° C (see Figure 4; note that E varies as velocity squared). Ide found that only a 2 percent reduction in the modulus resulted from heating to 100° C.[16] Griggs found no spalling or extension of cracks in photomicrographs after heating and cooling a granite block between 32° C and 142° C for about 20,000 cycles.[17] However, Ide's result at 100° C indicates that some microfractures would have formed, although they would have been undetectable at the magnification Griggs used. Hudec shows that water in the pores weakens rock, making thermal stresses more effective.[5] A property like elasticity could be used to reveal the extent of damage from thermal cracking. Modern ultrasonic, acoustic, or mechanical velocity-logging devices can be used to measure the expected decrease in elasticity and could be adapted to measure the weathering of monument stones.

Hooker and Duvall, in a quarry at Mount Airy, N.C., measured a 70-bar increase in stress in granodiorite resulting from a 25° C tem-

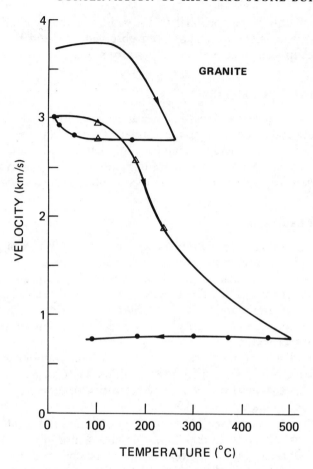

FIGURE 4 Irreversible change in longitudinal sonic velocity
(a measure of the elastic modulus E) of a Quincy granite sample
on heating to 270° C and then to 500° C (Ide).[16]

perature change between February and August (see Figure 5).[18] The
thermal stress equation is:

$$\sigma = \alpha E(T_1 - T_0)/(1 - v), \tag{5}$$

where σ is stress in kilobars; α is coefficient of expansion in reciprocal
degrees Celsius; E is elastic modulus in kb; T_1 and T_0 are final and
initial temperatures in degrees C; and v is Poisson's ratio, which can
be taken as about 0.25. Stress change per degree temperature rise was
measured at 3.1 bars per degree Celsius (Figure 5) and was calculated

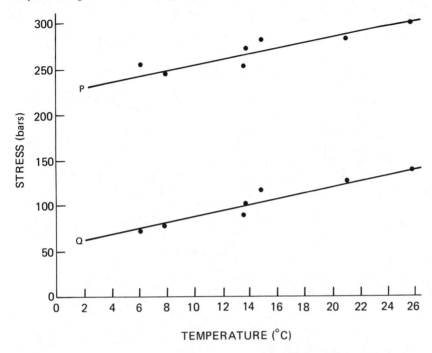

FIGURE 5 Increase in the two principal stresses, *P* and *Q*, (acting horizontally) from thermal expansion caused by a 25° C increase in temperature in granodiorite at Mount Airy, N.C. (Hooker and Duvall).[18]

to be 2.9 bars per degree Celsius. This was a very good corroboration for a superimposed effect on a rock under existing tectonic horizontal stresses (lines *P* and *Q* in Figure 5) of about 100 and 300 bars. In unconfined rock, a stress of 70 bars would approach the tensile strength.

The expansion of water in pores when it freezes is an important process in the deterioration of rock, but constraint is necessary for breakage to occur. Near the surface, where pores are large, the expanding ice can move into open space and not exert pressure. Deeper in the rock, expansion of the ice can be constrained by the tortuosity of pores and microcracks; if the tensile strength (indicated in Table 1 by modulus of rupture) is exceeded, the rock ruptures. At −10° C, constrained ice exerts 1 kb of pressure, much above the normal tensile strength of rocks. The pressure–temperature relations for ice are shown in Figure 6. Hudec found that "unsound" rocks having small pores are not as susceptible to deterioration by freezing as are "sound" massive rocks.[5]

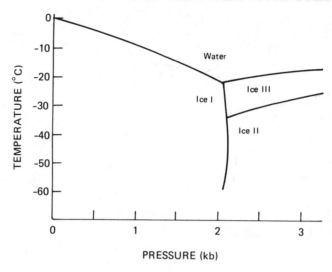

FIGURE 6   Part of the phase diagram of H_2O for liquid, and ice I,
II, and III (Winkler).[1]

Thermal Conductivity

Thermal conductivity, K, is the rate at which heat is conducted in
millicalories per second through a 1-cm^2 area down a temperature
gradient of 1° C over 1-cm length. In rocks, K is affected not only by
the mineral composition but also by the porosity, the degree of fluid
saturation, and heating. A compilation of thermal conductivities of
rocks is given by Robertson.[19] Quartz has a high K, averaging 18 mcal/
cm per second per degree Celsius; an increase of quartz content from
1 to 90 percent in a dense, dry, felsic sandstone increases K fivefold.
The K of most other minerals is a half to a fifth of that of quartz. As
porosity decreases from 40 percent to 1 percent—that is, as solidity
increases—the K of common rocks increases by a factor of two to
three. A temperature rise of 100° C causes a 10 percent reduction in
K in quartzose and ultramafic rocks; values for K for feldspar-rich and
basaltic rocks are much less affected by temperature.

Heat is conducted slowly in rocks; for instance, K is 20 to 50 times
larger in metals than in rocks. Thus stone, relatively, is an insulating
building material. The usefulness of K for stone deterioration problems
is in determining temperature changes in rock and analyzing the depth
of significant thermal expansion.

Diffusivity and Thermal Inertia

The diffusivity, k, can be calculated from thermal conductivity, K, density, ρ, and specific heat, C_p:

$$k = K/\rho C_p. \tag{6}$$

A similar parameter is thermal inertia, I, which is also used in analyzing cyclic thermal effects:

$$I = \sqrt{K\rho\, C_p}. \tag{7}$$

These parameters of heat conduction are useful in estimating changes in temperature with depth from a surface across which heat is transmitted.

Cyclic temperature changes, diurnal or seasonal, may be analyzed by the following equation:

$$T_x = T_s \exp[-x\sqrt{\pi/kP}], \tag{8}$$

where T_x and T_s are temperature variations at depth x and at the surface in degrees Celsius; x is depth in centimeters; k is diffusivity, which is essentially constant, in square centimeters per second; and P is period of temperature variation in seconds (1 day = 86,400 s). As the depth x increases, the exponential term becomes smaller; as the period P increases, it overwhelms k, and the exponential term increases. Therefore, long-period seasonal temperature changes penetrate deeper than diurnal ones, but the effect dies out rapidly with depth. Hooker and Duvall found only a 2° C change at 25-ft depth in granite for 25° C change at the surface; their calculated and observed temperatures through the 25-ft depth agreed closely.[18]

MECHANICAL PROPERTIES

Hardness

Mohs' scale of hardness (relative scratch hardness) for minerals is useful for many purposes. However, the Mohs hardness of rocks is difficult to determine and is too much affected by friability and surface texture to be really useful. Clearly, the ease of polishing monument stones is related to hardness, but stone deterioration is not obviously related to Mohs' scale. Impact and rebound hardness can be correlated with the strength of rock,[1] but their relations to rock decay are probably minimal.

Elasticity

The ratio of stress to strain is Young's modulus of elasticity, E; it is usually accepted as constant. Most rocks are brittle and behave elastically to an elastic limit, which is usually near the stress at which the rock fails. As noted under Thermal Expansion (above), microcracking changes the elasticity of a rock, so changes in elasticity may be used to detect increased porosity, permeability, and susceptibility to deterioration.

Compressive Strength

Walsh discusses compressive strength in detail in these proceedings, but a brief review may be in order. The unconfined, or uniaxial, strength of rock in compression is the maximum stress attained before the rock fails, usually by brittle rupture at strains of a few percent. Generally, igneous rocks, quartzite, and some slate are strong rocks; schist and marble are moderately strong; and the porous sedimentary rocks are weak. Limestone and rock salt, under respective confining pressures of 1 kb and 0.2 kb or higher, will deform plastically under differential stress above the elastic limit; they can deform by creep to 20 to 30 percent before large cracks form. The uniaxial strengths of porous sandstones, limestones, and shales depend on porosity; the strength of these rocks increases about threefold as porosity decreases from 35 to 1 percent. These rocks also show a small increase in strength as grain size decreases. In a feldspathic sandstone cemented by calcite, an increase in quartz content from 1 to 60 percent increases strength fourfold. The effect of water, especially if under pressure, is to weaken rocks; this effect and those of confining pressure and other physical conditions on graywacke and other rocks were reviewed by Robertson.[20]

Modulus of Rupture

The modulus of rupture of rock is measured by a simple bending test on an unconfined sample. It is approximately equal to the tensile strength, in that the sample fails by tension in the extended elements of the beam. Tests strictly of tension in rocks are quite difficult to do properly. As can be seen in Table 1, the modulus of rupture is one-third to one-tenth of the compressive strength. The modulus as a test of tensile strength is useful in applications involving failure in tension of building and monument stones because of thermal or mechanical microfracturing.

Friability

A rock or mineral is said to be friable if it crumbles naturally or is easily broken. Examples of friable rocks are soft or weakly cemented sandstones and shales. The friability of rocks can be considered a gradational mechanical property, which is perhaps best measured by an abrasion–hardness test. Such tests use a grinding powder and a lap or wheel applied to the specimen under a standard load. Relative values from such tests could be useful in detecting and monitoring the deterioration of building stones. No values of abrasion hardness are given in Table 1 because the values range widely among rocks and overlap from one rock to another. Friability depends on the strength of the weakest of the major mineral constituents and on the strength of the bonding between the mineral grains. Friability is low in dense, igneous rocks and high in porous, sedimentary rocks; it depends on the character of the intergranular bonds, from the weak bonds of a poor cement to the strong ionic bonds of silica tetrahedra.

OPTICAL PROPERTIES

Color

The colors and patterns of monument, facade, and other building stones are important for artistic reasons. However, aside from changes that indicate the extent of weathering, color is not important in stone deterioration.

Transmittance and Reflectance

Some stones like marble, travertine, and chalcedony are selected as facade stones for their transmission and reflection of light, because of their interesting layered patterns and colors, through thin slabs and from polished surfaces. Light reflected from mineral cleavage, twinning, or grain surfaces, as in calcite, labradorite, and mica, may indicate locations of cracks, which would enhance permeability and lead to decay.

ELECTRICAL PROPERTIES

The resistivity and dielectric strength of stones are not affected directly by decay, but both properties are influenced strongly by the permeability and saline-water content of the pores in the stones. Measurements of electrical properties could be used to estimate permeability and approximate water content. Brace found that the resistivity of a wide range of crystalline igneous and metamorphic rocks decreased as

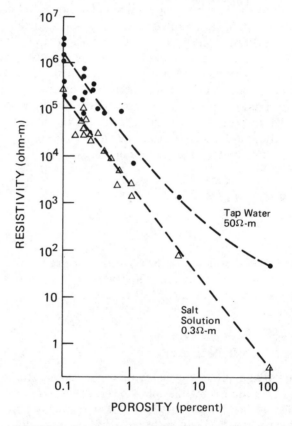

FIGURE 7 Decrease in resistivity of many crystalline igneous and metamorphic rocks with increase in porosity to about 5 percent, for saltwater and tap water saturating the samples under 4 k bar confining pressure (Brace).[21]

water content increased (see Figure 7).[21] Resistivities decreased from
10^6 ohm-m at ϕ = 0.1 percent to 10^2 ohm-m at ϕ = 5 percent in
samples saturated with saline water; resistivities were one-tenth as
high with tap water. If Brace's results were extrapolated to the higher
porosities (and corresponding water content) of sandstones, the resistivity would be about 1 ohm-m at ϕ = 40 percent. Thus, resistance
decreases rapidly as porosity and water content increase. Good resistivity measurements are easy to make, using four-probe geophysical
techniques, and can detect small changes in ϕ and water salinity and
saturation.

MAGNETIC PROPERTIES

The magnetic susceptibility and remanant magnetism of rocks are
closely tied to the magnetite content. However, magnetite is a minor
constituent of rocks and without importance to decay processes in
stones.

BUILDING AND MONUMENT STONES

Winkler reviewed the specifications for stones of the American Society
for Testing and Materials.[1] The ASTM tests are guidelines to proper
selection of stone for specific uses, although they need to be brought
up to date. In addition, the comprehensive works of Bowles[2] and Barton[7]
provide very useful descriptions of the properties of building and monument stones and of criteria for selection. The tests and descriptions
of criteria for selection inherently provide information on susceptibility to deterioration, but the physical and chemical mechanisms of
deterioration need thorough study. Where structures and monuments
have already deteriorated, the physical properties of their stones will
need study in properly diagnosing and solving problems.

Physical properties are measured on small specimens of stones, but
minor and subtle features of rock in place can be of overriding importance. Merrill reported on a new firm that started up an abandoned
but formerly successful quarry and lost nearly $1 million because the
new operators failed to observe imperceptible defects in the rock in
the new quarrying zone.[3] He said that, as a consultant, if he were
restricted to either field examinations or laboratory tests, he unhesi-
tatingly declares that, with good natural outcrops or quarry openings
of long standing, he would choose the field examination, no matter
how elaborate the other tests might be. At the time of writing, he was
probably correct, but today, presumably, careful sampling and complete testing of physical properties can detect small but critical differ-

ences in the characteristics of building stones and thus reinforce the field examinations.

Igneous Rocks

Granite

In geologic usage the name "granite" refers to rocks of various origins, a range including felsic igneous and metamorphic rocks that vary considerably in mineral composition. These rocks usually are dense and range in grain size from fine and equigranular, through medium and equigranular or porphyritic, to coarsely granular. Porosity and permeability are usually low, and the granite has high resistance to weathering and corrosion unless it is highly jointed, microfractured, or foliated.

Granites range in jointing from those having no definite rift, like the granites at Charlotte, North Carolina, and Vinalhaven, Maine, which can occur in blocks 90 m by 6 m by 3 m (300 ft by 20 ft by 10 ft), to a Wisconsin granite having joints at 20 cm (8 in.) spacing, too close for a building stone. Gneissic granite is strong perpendicular to its foliation but can be split into slabs for curbing and paving stones. Incipient joints, which are actually planes of microfracturing, occur in the granite of Essex County, New York; although the granite is acceptable for buildings, the incipient joints would open up on prolonged exposure and deface a monument. Calcareous layers in mica schists continue into contiguous massive granite gneisses in Vermont and Maryland and in time would be sources of deterioration. Gabbro bodies are seldom quarried in the United States because the rock is hard and difficult to work.

Rhyolite and Andesite

Rhyolite and andesite are volcanic rocks that occur as massive rock and as porous or welded tuff. The porous tuff is usually poorly consolidated, has bedding partings in places, and has a density less than 2 g/cm^3. It is subject to permeation by rainwater but drains well and has the virtues of easy workability and good standing strength; the effects of frost can be severe. The welded tuff is very hard and difficult to work. Indurated rhyolite and andesite do not take a polish and are seldom used for buildings in the United States. They have been used in the past in Europe, however, because of their easy workability.

Basalt

The dense varieties of basalt are relatively impervious to water, but they are hard and lack a rift. Columnar jointing is found in certain basalt flows, and the columns have been used in a few buildings. Basalt in massive flows is dark and does not take a polish, so it has little aesthetic appeal. Jointed basalt will have crack permeability, but vesicular basalt may not be permeable owing to isolation of vesicles.

Metamorphic Rocks

Quartzite

The strong, dense quartzites are usually cemented by silica, are fine grained, and are almost impervious to moisture. The Dakota quartzite is a good example. It takes a fine polish, although only after considerable grinding; it is unique in that it has almost perfect rift and grain cleavages. These properties make the stone desirable for ornamental as well as building uses. Silica-cemented, fine-grained quartzites do not deteriorate, but if shaley layers or close jointing occur, they will constitute planes of weakness and high permeability.

Marble

Both calcitic and dolomitic marble are massive rocks but commonly have moderate intrinsic permeability. In fact, moderate friability sometimes develops after only a few years of weathering, especially in coarse-grained marble. Tremolite laths, which are ubiquitous in marble, weather out and leave pocks. Marble occurs in a variety of colors and polishes well; however, it is usually jointed and fairly permeable and therefore can be subject to rapid chemical decay. Blasting and rough mechanical treatment create microcracks easily in marble and are avoided in good quarrying practice.

Slate

The obvious cleavage of fissile slate provides permeability for water penetration. Good roofing slate is fine grained, smooth, and tough. The quarried slabs have good bonding and low permeability across the unbroken cleavage and are quite resistant to deterioration.

Sedimentary Rocks

Sandstone

The type of intergranular cement determines the physical characteristics of sandstones. There are four common cements—silica, limonite, calcite, and clay minerals. Silica-cemented sandstone, even though porous, may resemble a quartzite in its high hardness, strength, and resistance to decay. Sandstone cemented by limonite is soft to work; in a fairly dry climate it will season to a harder, stronger rock, resistant to weathering and chemical disintegration. Calcitic cement is susceptible to the same kinds of chemical decay that affect limestone, and sandstone containing it may be greatly weakened. Clayey cement absorbs water, and sandstone containing it is easily broken, either by freezing or because clay minerals form poor intergranular bonds.

The reddish-brown, porous sandstone from Seneca, Maryland, used in the Smithsonian building in Washington, D.C., is limonite-cemented and has stood up reasonably well. The sandstone in Potsdam, New York, has both limonite and silica cement, and so it is soft to work and also holds up against deterioration. The Berea, Ohio, grit is fairly porous and easily worked; it has very little cement, probably silica. The material is cohesive, but slightly friable, and is used for grindstones because the grit contains none of the other cements that would glaze the surface and stop the cutting action.

Inhomogeneities resulting from interbedding are important to observe in checking sandstone formations for use as building stone. Adjacent layers may differ considerably in type of cement or plagioclase and mica content, or shale beds may be intercalated; if so, the sandstone, whether soft or hard, may be unusable for buildings. Also, the bedding may be too thin for usable blocks. Such variation can be very subtle, and close observation is needed. Because the porosity of commercial sandstone for building use ranges from 2 to 15 percent,[7] the permeability will be fairly high and the stone will stand up to decay only if interbedded layers are not permeable. Quarrying of sandstone is often stopped in winter because water deep inside fresh blocks would freeze and split them, and freezing of near-surface water could cause spalling.

Limestone

The relatively easy workability of limestone makes it a favored stone for construction and monuments. However, almost all limestones, even those of low porosity, have relatively high intrinsic and micro-crack and bedding permeability. Thus they are susceptible to weathering by permeating water and gases and especially by the well-known process involving conversion of sulfur dioxide to gypsum. Bedford, Indiana, limestone is soft but moderately strong; it has no rift and can be worked in any direction, so it is a much-used building stone. Dense oolitic limestones are commonly varicolored, compact, and easily polished, so they are used as veneer or ornamental stone. Normal limestone is usually impure, containing quartz, mica, clay, other silicate minerals, and carbon; shaley layers along the bedding can form partings as a result of weathering. In fossiliferous limestone in Kansas, the space around the fossils is not filled, and cellular breakage occurs. The soft limestone at Caen, France, is easily carved and, being moderately strong, is widely used for buildings in Europe. However, it deteriorates rapidly in the more severe U.S. climate because of its relatively high permeability.

Shale

Shale is inherently friable in that it is not lithified well enough to resist abrasion. Shale has very low intrinsic permeability because of its clay mineral content, but some shales have pronounced bedding planes and jointing, which provide permeable channels if the shale is under very low confining pressure. Invasion of shale by water often results in almost complete disintegration; adobe and mud for walls, which react to water by disintegration, are essentially shalelike in mineral composition.

Soapstone, Travertine, and Serpentinite

Soapstone, travertine, and serpentinite are relatively soft and take a good polish; they are used for ornaments, statuary, or facades. They rarely occur in large blocks. Travertine usually is soft just after quarrying and becomes hard on standing; of course, it is porous and subject to the deterioration characteristic of such calcitic stones. Verde antique is ornamental serpentinite and usually consists of white calcite veins running through the variegated green serpentine. It is used as a veneer stone. The calcite, of course, can be corroded by atmospheric moisture and gases.

CONCLUDING REMARKS

Knowledge of physical properties can be useful both for initial selection and for diagnosis of the deterioration processes of building stones. We need not only laboratory measurements of physical properties, but also observations on the rock in place. For example, laboratory measurement of the strength of rock is not an adequate index of the stability of stone under weathering or other decay processes. Knowledge of lack of homogeneity in macroscopic to microscopic features needs to be obtained by quarry-site inspection and by microscopic observations of petrographic and textural discontinuities. Once identified, inhomogeneities can be tested in the laboratory on carefully selected samples; there are many examples of such testing.[1,5,7,8,9,10,11]

In stone to be used for buildings or monuments, such physical properties as rift (cleavage parallel to foliation or bedding) and grain (cleavage perpendicular to rift) should be identified. As has been repeatedly mentioned above, veins, layers, or cements of calcite, clay, talc, mica, or shale, whether thick or thin, can be expected to weather out or be corroded and, together with bedding and cross joints, can increase permeability. Pore size and shape in stone can vary from one part of a quarry to another, even within short distances. Such variation can influence the effective porosity or permeability, and the permeability may vary along and across bedding or foliation in a single quarry. Thus, induration, foliation, microfracturing, variation in mineral composition among or within layers, and jointing at small to large intervals in rock in place can be more significant than laboratory tests in determining the susceptibility of building and monument stones to deterioration. These characteristics also are important in diagnosing deterioration processes affecting stones in use.

Quarrying methods can affect the durability of stones. Blasting can create cracks that become permeable channels for water. The same effect can be produced by imperfect splitting along rift and grain. It can also result from zones of small-scale microfracturing, which can form when existing tectonic stresses are concentrated by the quarrying operation until they exceed the strength of the rock and it fails.

For those analyzing deterioration processes in particular stones, geophysical techniques could provide useful measurements of the physical condition to supplement laboratory tests of physical properties. Geophysical exploration techniques based on electrical resistivity and acoustic velocity can probably be helpful in diagnosing stones undergoing corrosion or decay. Resistivity varies with water content and salinity and so would be sensitive to increases in permeability and porosity resulting from dissolution or microfracturing. Ultrasonic-wave-velocity methods can be used on a small scale (to 10 cm) to detect spalling or microfracturing from thermal or other stresses in the stones, but acoustic-wave-velocity measurements could be used for deeper penetration (to 100 m). Hudec found that water saturation decreased velocity in "sound" rocks and increased velocity in "unsound" rocks (which are more susceptible to weathering),[5] so saturation must be accounted for in interpreting velocity studies.

In general, to obtain a fuller explanation of each deterioration process in building stone, collaboration will be needed among the following people: the preservationist, who knows where these processes take place, what stones to study because of their architectural and historic significance, and what remedies have been tried; the geologist, who knows the origin and mineral content of the stone and its probable geologic inhomogeneities; the specialist in rock mechanics, who knows the measurement and the significance of physical properties; the geophysicist, who knows exploration techniques and their application to characterizing the extent of stone decay; and the geochemist, who understands the chemistry of weathering and knows what analytical techniques can be used to explain the deterioration process. With fuller explanation will come knowledge of what physical and chemical properties to measure and how to measure them, leading to more satisfactory decisions on remedial measures.

REFERENCES

1. Winkler, E.M., 1973, *Stone: Properties, Durability in Man's Environment*, Springer-Verlag, New York.

2. Bowles, O., 1934, *The Stone Industries*, McGraw-Hill, New York.

3. Merrill, G.P., 1903, *Stones for Building and Decoration*, John Wiley, New York.

4. Schaffer, R.J., 1932, *The Weathering of Natural Building Stones*, Harrison and Sons, London.

5. Winkler, E.M., ed., 1978, Decay and preservation of stone, Geol. Soc. Amer., *Eng. Geol. Case Histories* No. 11, 104.

6. Moen, W.S., 1967, *Building stone of Washington*, Washington Div. Mines & Geol., Bull. 55.

7. Barton, W.R., 1968, *Dimension Stone*, U.S. Bur. Mines Infor. Circ. 8391.

8. Blair, B.E., 1955, 1956, *Physical Properties of Mine Rock*, Parts III, IV, U.S. Bur. Mines Rep. Inv. 5130 and 5244.

9. Blair, B.E., 1956, *Physical Properties of Mine Rock*, Part IV, U.S. Bur. Mines Rep. Inv. 5244.

10. Clark, S.P., Jr., 1966, *Handbook of Physical Constants*, Geol. Soc. Am. Mem. 97.

11. Windes, S.L., 1950, *Physical Properties of Mine Rock*, Part II, U.S. Bur. Mines Rep. Inv. 4727.

12. Brace, W.F., 1980, Permeability of crystalline and argillaceous rocks, *Int. J. Rock Mech. Min. Sci.* v. 17, no. 5, p. 241–252.

13. Morrow, C., Lochner, D., Moore, D., and Byerlee, J.D., 1981, Permeability of granite in a temperature gradient (abstract), *EOS*, v. 61, no. 52, p. 1238.

14. Blatt, H., Middleton, G., and Murray, D., 1980, *Origin of Sedimentary Rocks*, Prentice-Hall, Englewood Cliffs, N.J.

15. Torraca, G., 1976, Brick, adobe, stone, and architectural ceramics: Deterioration processes and conservation practices, in Proc. North Amer. Int. Reg. Conference, 1972, *Preservation and Conservation: Principles and Practices*, S. Timmons, ed., Smithsonian Inst. Press, Washington., D.C., pp. 143–165.

16. Ide, J.M., 1937, The velocity of sound in rocks and glasses as a function of temperature, *J. Geol.*, v. 45, no. 7, pp. 689–716.

17. Griggs, D.T., 1936, The factor of fatigue in rock exfoliation, *Jour. Geol.*, v. 44, pp. 783–796.

18. Hooker, V.E., and Duvall, W.I., 1971, *In Situ Rock Temperature: Stress Investigations in Rock Quarries*, U.S. Bur. Mines Rep. Inv. 7589.

19. Robertson, E.C., 1979, *Thermal Conductivities of Rocks*, U.S. Geological Survey Open-File Report 79–356.

20. Robertson, E.C., 1972, Strength of metamorphosed graywacke and other rocks, in *The Nature of the Solid Earth*, E.C. Robertson, ed., McGraw-Hill, New York, p. 631–659.

21. Brace, W.F., 1971, Resistivity of saturated crustal rocks to 40 km based on laboratory measurements, in *The Structure and Physical Properties of the Earth's Crust*, J.G. Heacock, ed., American Geophysical Union Monograph 14, p. 243–255.

Deformation and Fracture of Rock

JOSEPH B. WALSH

Rock is not elastic in the usual sense of the word. Cracks, which are found in nearly all rocks, close under compressive stresses and make the sample stiffer. One side of a crack slips over the opposite side under differential stress, and energy is lost to friction. As a consequence, stress–strain curves are non-linear, and hysteresis occurs in complete loading–unloading cycles. The tensile strength of rock is much less than the compressive strength. Strength is not affected appreciably by moderate changes in temperature or by the rate at which the load is applied. The cracks and pores in rock form a continuous, interconnected network, and rock under natural conditions is usually saturated with fluid. High pressure or chemical activity of these pore fluids decreases strength. The strength of rock increases dramatically as confining pressure increases.

The deformation of most materials used in construction is described for each by a stress–strain curve showing how strain (deformation per unit length) varies with stress (load per unit area). The stress–strain curve is approximately linear with most solid materials for small changes in stress, and the well-known Hooke's law applies. Materials for which

Joseph B. Walsh *is Senior Research Scientist, Department of Earth and Planetary Sciences, Massachusetts Institute of Technology, Cambridge.*

This work was supported by the U.S. Geological Survey under Contract No. 14-08-0001-18212 and by the Army Research Office, Durham, under Contract No. DAAG29-79-C-0032.

stress is proportional to strain until fracture occurs are called linearly elastic; the stress–strain curve for glass, which is such a material, is shown in Figure 1. Elastic materials have the characteristic that strain is a single-valued function of stress; Figure 1 shows that the strain at any specific stress when the load is being increased is the same as the strain when the load is being decreased. Some materials (rubber, for example) are described as nonlinearly elastic; in such cases, strain is a single-valued function of stress, but the stress–strain curve is not a straight line.

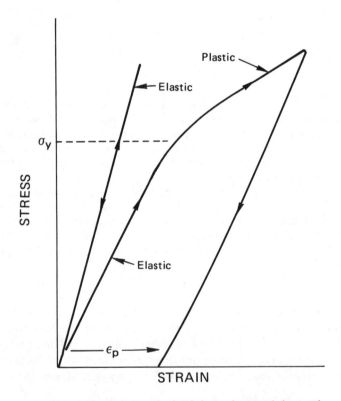

FIGURE 1   Deformation can be divided into elastic and plastic. The stress–strain curve for an elastic material like glass is the same during the loading and unloading portions of a complete cycle. Ductile materials like many metals may be elastic until a yield stress, σ_y, is reached. Plastic flow occurs at higher stresses, and the complete curve describes a hysteresis loop with permanent strain ε_p.

The deformation of a large class of materials (most metals, for example) is elastic for stresses less than the yield stress, and further increases in load result in plastic deformation. In plastic deformation the movement of dislocations allows whole planes of atoms to translate relative to the adjacent plane, with the consequence that relatively large strains occur for small changes in stress. The atomic planes do not translate in the opposite sense when the load is lowered, and so the stress–strain curve in the loading cycle is not the same as that during the unloading cycle. A stress–strain curve for a material deformed into the plastic region is shown in Figure 1; note that the specimen has undergone a permanent change in length.

The type of deformation that a sample undergoes depends on the type of stresses applied, as well as on their magnitudes. The stresses acting on a body can be considered to be the sum of two stress systems—the hydrostatic component and the shear or deviatoric component. Most materials, even ductile materials, are found to behave elastically when only hydrostatic stresses are applied. Elastic materials are those that behave elastically under both hydrostatic and deviatoric stresses, whereas ductile materials undergo plastic deformation at deviatoric stresses greater than the yield stress.

When the load is increased to a sufficiently high level, a specimen eventually fractures. Fracture also can be divided into subcategories that include most engineering materials. "Brittle" failure is said to have occurred when the fractured pieces can be fitted back together—that is, when the sample has undergone a negligible amount of permanent deformation before and during the fracture process. Glass, of course, is a common example of a brittle material. A fracture is called "ductile" when the sample undergoes an appreciable amount of plastic flow before fracture. A relatively large amount of energy is absorbed by materials that fail in this way, and so such materials are used where toughness is important.

The type of fracture that occurs does not depend completely on the material involved. The configuration of the specimen is important: The presence of cracks and sharp notches tends to inhibit ductility and, in extreme cases, ductile material can fail by brittle fracture. The loading rate and the temperature can also affect the type of fracture that occurs, with high loading rates and low temperatures enhancing the likelihood of brittle failure.

This brief discussion of deformation and fracture is intended to provide only a framework for describing the behavior of rock. Deformation and fracture are well-developed scientific fields, and the paragraphs here do not provide even an elementary introduction to these complex processes. A comprehensive introduction to the subject can be found in *Mechanical Behavior of Materials*.[1]

PROPERTIES OF ROCK

The pressures and temperatures of interest in building construction are at the low end of the broad range of conditions that has been studied in geology. Even in this restricted range, rock is found to have unusual properties. The deformation cannot be characterized as either elastic or plastic, and fracture involves elements of both brittle and ductile behavior. Most of the minerals that make up rock are elastic, hard, and brittle. Rock-forming minerals are hard and brittle because of the low symmetry of their crystallographic structures. Dislocations cannot move easily in these complicated structures, so plastic flow is almost completely inhibited. One common mineral, calcite, of which marble and limestone are composed, is exceptional in that it can be deformed plastically along one crystallographic plane. However, calcite crystals of most orientations cleave when stressed, and the overall behavior must be classified as brittle.

The question, then, is how can rocks that are composed of elastic, brittle crystals exhibit anything but elastic, brittle behavior?

DEFORMATION OF ROCK

The measurement of the compressibility of a sample of rock is a standard test that provides basic information about the rock's internal structure. Typically, a specimen is sheathed with an impermeable jacket and put in a pressure vessel. Strain gauges on the jacket, or various other devices, are used to measure the decrease, ΔV, in the volume, V, of the sample as the pressure, p, in the vessel is increased. A curve of volumetric strain $(\Delta V/V)$ as a function of pressure is developed from these data, and compressibility, β, is calculated from the inverse slope $(\Delta V/Vp)$ of the curve.

An example of a volumetric strain–pressure curve for a granite from Westerly, R.I., is shown in Figure 2. Note that the curve is nonlinear. At low pressure, the slope is relatively low; the slope increases with increasing pressure until at a pressure of approximately 2 kbar (about 30,000 psi) the curve becomes linear. Note also that the curve relating strain and pressure when pressure is decreased is virtually the same as the curve when pressure is increased. Hysteresis is negligible, and behavior is nonlinearly elastic.

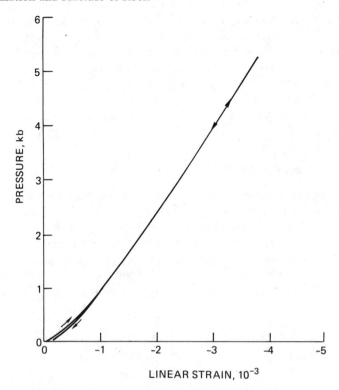

FIGURE 2 Volumetric strain of a sample of Westerly granite as a function of pressure. Note that this rock is nonlinearly elastic under hydrostatic pressure.[3]

The generally accepted explanation for this behavior, which is observed for virtually all types of rock, is that rock contains cracks. At low pressure the cracks are open, and the rock is compliant. Cracks close as the pressure is increased, and the sample becomes stiffer. At sufficiently high pressures, typically 2 kbar, all cracks are closed, and the sample behaves like its constituent minerals, which are linearly elastic.

The cracks found in rock arise from several natural causes. The cooling of a rock after it has solidified creates thermal stresses that are sufficient to fracture the crystalline framework. Likewise, microcracking can result from the decrease in stress that a sample experiences when it moves from its origin at depth to the surface or from the increases in stress that occur during tectonic deformation.

The exsolution of gases during solidification or the imperfect sintering of grains, for example, also increases porosity. Pores are cavities that are not cracklike (i.e., one dimension is not much smaller than the others) and their effect on deformation is entirely different. Figure 3a shows a rock sample, containing both cracks and pores, under low pressure. When sufficiently high pressure is applied, as in Figure 3b, the cracks close, but the pores do not. Therefore, the nonlinearity in the pressure–strain curve is caused by the cracks, not by the pores.

Effects of Cracks and Pores

Some progress has been made in evaluating quantitatively the relative effects of cracks and pores on the elastic properties of rock (for a recent review, see Walsh).[3] Using techniques in the theory of elasticity, we find that cracks have a very much greater effect on compliance than do pores of equivalent volume. For example, spherical pores that constitute a few percent of the volume of a sample cause an increase in compressibility of a few percent. A crack is found theoretically to have

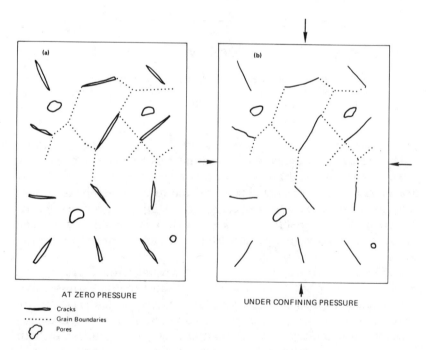

AT ZERO PRESSURE

UNDER CONFINING PRESSURE

— Cracks
···· Grain Boundaries
Pores

FIGURE 3 A schematic description of the effect of confining pressure on cracks and pores. Note that cracks close under pressure, whereas pores do not.

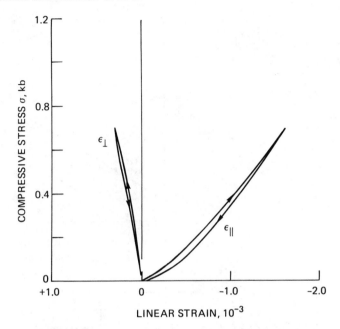

FIGURE 4   Longitudinal and lateral strain of Westerly granite under uniaxial compression.[2] Note that the stress–strain curves are nonlinear and nonelastic.

nearly the same effect on compressibility as a pore of the same diameter. The porosity associated with such a crack is negligibly small compared with the pore, of course. A rule of thumb is that the effect of cracks on elastic properties depends on the total surface area of the crack phase (and is independent of its volume), whereas the effect of pores is in direct proportion to the volume of the pore phase. The compressibility of the Westerly granite sample in Figure 2, for example, is 8.3 mbar^{-1} at atmospheric pressure and 1.87 mbar^{-1} at high pressure. Cracks in this rock, which account for a porosity of only 0.5 percent, have increased compressibility by a factor of nearly 6. The effect of the pores in the sample, which account for a porosity of approximately 1 percent, is negligible.

Cracks also have a pronounced effect on the deformation of rock when stresses other than hydrostatic pressure are applied. Figure 4 shows the longitudinal and lateral strain caused by increasing the uniaxial compressive load on a cylindrical sample of Westerly granite. Compare the longitudinal strain–stress curve in Figure 4 with the volumetric strain–pressure curve in Figure 2. The slope of both curves

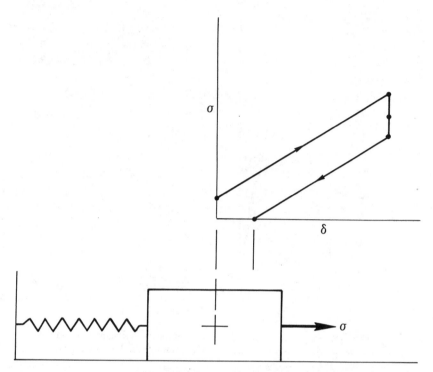

FIGURE 5 A schematic description of the effect of sliding of crack faces against friction. Note that slip does not occur immediately as the stress is lowered. A hysteresis loop is formed, and the system undergoes permanent displacement.

increases with increasing stress. The reason for this is the same in both cases: Increasing compressive stress or pressure causes cracks to close, and the specimen becomes stiffer. Eventually all, or nearly all, cracks close, and strain is a linear function of stress.

However, note in Figure 4 that the curve when the applied stress is being decreased is not the same as when the stress is being increased, in contrast to the response to hydrostatic-pressure changes in Figure 2. This behavior is typical of most rocks—that is, rocks are nonlinearly elastic under hydrostatic-pressure changes and are nonelastic under changes in deviatoric stress.

Cracks are responsible for the nonelasticity as well as for the non-linearity of rocks. Cracks under changes in hydrostatic pressure merely open and close. However, one side of a suitably oriented crack can slide relative to the other side when the applied stresses contain a deviatoric component. This relative motion means that the longitudinal strain of the sample is somewhat greater than the strain from compression of the solid matrix; as a consequence, Young's modulus (the slope $d\sigma/d\varepsilon_\parallel$) of the ascending portion of the curve in Figure 4 is less than Young's modulus for the mineral components.

The process is illustrated schematically in Figure 5. Slip between the faces of a microcrack in rock is physically similar to sliding a block against friction on a plane. The spring in Figure 5 represents the elastic element in the deformation owing to the mineral grains. Note that the block does not immediately begin to move in the opposite sense when the load is decreased, much as a door forced against a wedge must be yanked in the opposite direction to dislodge it. Consider the situation at the highest stresses in Figure 5, where all cracks are closed and those cracks that can slide are sliding. Using the analogy of a block sliding on a plane, we see that all these cracks will be jammed when the load initially is lowered. The inference is that the initial slope of the unloading curve in Figure 4 represents the response of only the mineral components. The response of the minerals in the rock under conditions in which cracks do not affect behavior can be measured in other ways. We find that, indeed, the initial slope of the unloading curve in Figure 4 represents an adequate approximation of the behavior of intact rock: Young's modulus for intact rock is 730 kbar, and the value from Figure 4 is 710 kbar.

Effects of Water and Temperature

The deformation of rock is influenced by factors in addition to stress. The porosity in most rocks forms a phase that is largely interconnected. As a result, rock in a wet environment becomes permeated with water. The response to uniaxial stress of two sandstone samples, one of them saturated with water and the other one dry, is shown in Figure 6. The wet sample is more compliant than the dry one. The analogy of a block sliding on a plain offers a plausible explanation for this behavior: Water lubricates the microcracks, slip on cracks is facilitated, and consequently compliance is increased.

Relatively small changes in temperature do not affect the deformational characteristics of rock appreciably as long as the temperature variations do not change the degree of fluid saturation. The elastic properties of the constituent minerals do not vary with temperature to an appreciable extent until the temperature reaches a level of approximately half the melting temperature (in kelvins); these temperatures are well above those that most building stone is subjected to

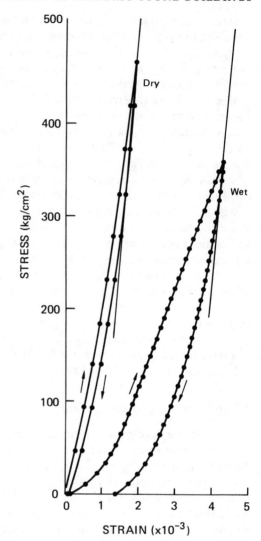

FIGURE 6 Stress–strain curves for dry and wet sandstone samples. The wet sandstone is more compliant than the dry.[4]

under normal circumstances. Temperature changes of 20° C to 30° C, however, are sufficient to cause measurable thermal cracking in competent granites,[4] thereby changing their compliance. Presumably, however, most building materials already will have been subjected to seasonal changes of that magnitude during their history. Although excursions in temperature beyond the range that the sample has experienced may cause cracking, repeated cycling to some given level does not appear to have any cumulative effect.

FRACTURE OF ROCK

A specimen eventually fractures when the stress acting on it is raised to sufficiently high levels. The stress at which fracture occurs depends not only on the type of rock and its previous history but also on factors such as the types of stresses that are applied, the degree of saturation and pressure of the pore fluid, and the rate at which the load is applied. Temperature change over the range experienced in building construction is not a factor of major importance in the discussion here.

Fracture research has benefited greatly in recent years from the development of the scanning electron microscope. Photographs of the interior of two samples of Westerly granite that were taken on a scanning electron microscope are shown in Figure 7. Figure 7a shows cracks that were produced by the applied stress at a level before the specimen was completely fractured. The photomicrograph in Figure 7b illustrates the damage produced by the applied stress approximately at the instant of fracture. Note that fracture at the microscopic scale is a brittle process. The crack surface in Figure 7a could be rejoined with virtually no gaps remaining; the microshards in Figure 7b are angular, with none of the rounded or stretched features typical of ductile fracture, and presumably the tiny blocks could be rejoined, with sufficient patience, to re-create the original state.

The cracks in these photographs have fractured individual mineral crystals to produce the fracture pattern that we see. The fracture strength of the rock, one might suppose, must be directly related to the strengths of the minerals involved. The strengths of single crystals have been investigated theoretically using several different approaches. In all cases the fracture strength, σ_t, is given approximately by the relationship:

$$\sigma_t \simeq E/10, \tag{1}$$

where E is Young's modulus. Young's modulus for quartz, for example, to an order of magnitude, is 10^6 bars,[7] so the theoretical strength is 10^5 bars. The fracture strength of a quartz crystal is less than 10^4 bars in compression and less than 10^3 bars in tension, and the values for quartzite (polycrystalline quartz rock) are lower by a factor of 10.[8] Why is the measured strength so low?

Griffith was the first to explain the discrepancy.[9,10] He reasoned that no material was perfect and that flaws in the form of tiny cracks could

FIGURE 7a A photomicrograph of Westerly granite at a stress well below the fracture strength.[6] Note that cracks have begun to extend and that the fracture is brittle.

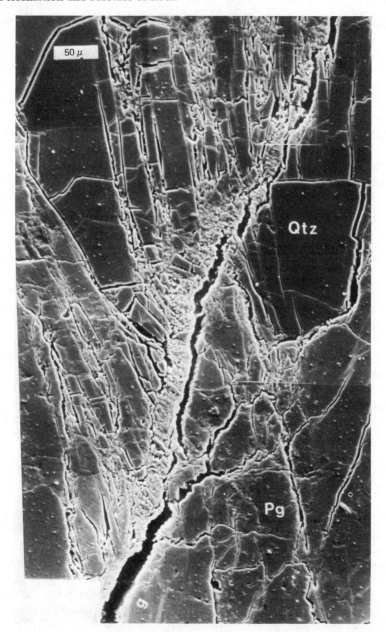

FIGURE 7b   A photomicrograph of Westerly granite on the verge of fracture.[6]
Note that fracture on the microscale is brittle, although the stress–strain curve
has the characteristics of a ductile material.

be found in solid samples of any size. Using a theoretical analysis of the stresses around cracks, he showed that the maximum stress, which is found at the tip of a crack, is very much greater than the applied stress. The stress, σ_f, required to break a specimen is therefore very much less than the theoretical strength, σ_t, given by equation 1, needed to extend the crack. Griffith's analysis showed that, in fact, the actual strength should be lower for materials having longer cracks. In an elegant experiment using samples of smaller and smaller size, he extrapolated to a sample of zero size (and zero crack length) and found the theoretical strength to be in agreement with the value given by equation 1.

Griffith's first analysis was restricted to fracture under tensile loading, but he generalized the results in his second analysis to loading under compressive states of stress. Griffith's theory showed that the strength of brittle materials like rock in compression should be about 10 times the tensile strength. Further, his theory showed that the compressive strength should rise dramatically when confining pressure is superposed. These theoretical predictions have been verified, at least in a general way, for rock and masonry materials by subsequent experimentation.

Griffith's theory, although it describes in broad outline how the strength of rock depends on the types of stresses applied, does not accurately describe the fracture process itself.[6,12] Griffith visualized that fracture in compression occurs as it does in tension: Cracks are inactive until the fracture stress is reached, when the largest crack grows across the specimen, separating it into two pieces. The stress–strain curve for a rock sample in Figure 8 shows that this model cannot be correct.

Figure 8 shows that the slope of the stress–strain curve first increases as cracks close and then begins to decrease. The decrease in slope is due to the growth of cracks in the sample. A careful examination of stress–strain curves like that in Figure 8 shows that cracks begin to extend when the applied stress is about one-third of the fracture strength. The number and length of cracks increase as the applied load is increased, and gradually the specimen is weakened until finally a fault is formed, fracturing the sample. Fracture in compression, then, is a complex process involving the growth and coalescence of many cracks.

The total nonelastic deformation that can be attributed to the growth of cracks (and also the fracture strength) is found to increase with increasing confining pressure. At sufficiently high confining pressures, the stress–strain curves and the shape of the specimen cannot be distinguished from those for truly ductile materials, although the mech-

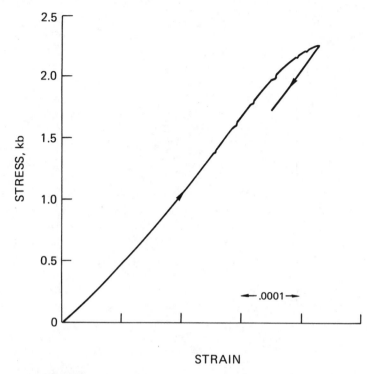

FIGURE 8   A complete stress–strain curve for Westerly granite under uniaxial compression.

anism on the microscale in one is brittle failure and, in the other, plastic flow.

Effects of Fluids

The pressure of fluid in the pores in the rock has a direct effect on fracture strength. Theoretical analysis and laboratory experiments show that pore pressure can be handled effectively by using the "law of effective stress." The effective stress for fracture is given by the applied stress less the pore pressure. The law of effective stress requires that all combinations of applied stress and pore pressure that produce the same effective stress must have the same effect on fracture. As a consequence, the fracture strength for any value of pore pressure and confining pressure can be found once the fracture strength has been established for one set of conditions. An example of how strength and effective stress are related can be found in Figure 9.

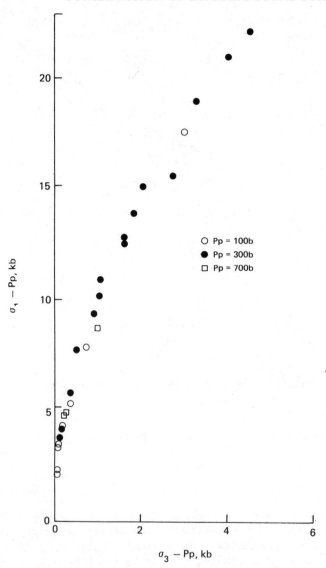

FIGURE 9 Compressive strengths for samples with different pore pressures all fall on the same curve when plotted in terms of effective stress.[11]

Pore fluids can also affect strength indirectly. Some fluids, including water, have a corrosive effect on rock-forming minerals. These fluids weaken rock in two ways. The corrosive action in situations where fluid can circulate through the pore and crack network gradually erodes the passages and weakens the matrix. This mechanism is not particularly important when fluids of natural origin are involved, except where the fluids are unusually corrosive to the rock minerals or the fluid is unusually hot. Fluids are also found to affect strength through a mechanism involving surface tension. The fracture strength of samples of a quartzitic shale is shown in Figure 10. Saturating rock with a fluid having low surface tension increases its load-carrying capacity.

Another indirect way in which fluids affect strength involves the permeability of the rock, the size of the sample, and the rate at which the load is applied. Consider a large sample of a relatively impermeable rock that is saturated with water, and assume that the load in the sample is increased very quickly. Fluid in the rock, though it has access to the outside, is trapped in the pores and cracks because of the low permeability and the long path. The pore pressure rises, and the strength decreases because of the law of effective pressure. Appreciable decreases in strength have been observed in the laboratory on samples of granite as small as a few centimeters.[10]

Size of Samples

Griffith's theory shows that specimens with long cracks have low strength. This suggests that large samples may be weaker than small samples because the probability that a sample contains a long crack is greater for larger samples. The effect of size, which is a matter of concern in manufacturing engineering practice where the range of sizes is relatively small, could be a major factor where the strength of rock is important because the range of interest can be enormous. A number of studies on specimens having a range of sizes (of the order of 1 to 10 cm) suitable for experiments in the laboratory have demonstrated that bigger samples of rock are indeed weaker than smaller ones. These studies show that, in general, the weaker the rock, the larger the effect of size on strength. The relationship between size and strength determined in these studies is usually expressed in the form

$$\sigma_f \sim (\text{size})^{-a}, \tag{2}$$

where a for strong rocks like competent marbles and granites is near 0.1 and, for a weaker rock like coal, is 0.5.

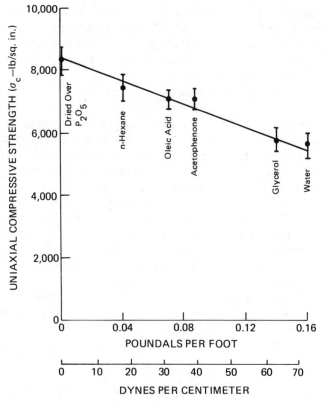

FIGURE 10 Compressive strengths of samples saturated with various fluids. Note that strength is high for fluids with low surface tension.[9]

The range of sizes used in these laboratory studies is barely in the range of interest in building construction. Studies in the field on very large specimens are rare because they are expensive and very difficult to do. The very limited number of observations that have been made on the strength of large samples suggests that equation 2 is valid only for samples less than a critical size, and the strength of larger samples is nearly the same. An example is shown in Figure 11.

Increasing the confining pressure tends to mitigate the effect of size. Griffith's theory shows that weak rocks have large cracks, and the above discussion of rock deformation demonstrates that cracks close

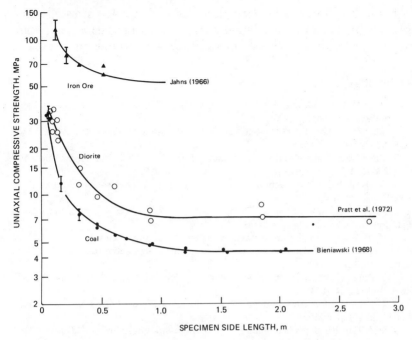

FIGURE 11 The compressive strength of rock is found to be smaller for larger samples.[15,16,17]

under compressive stress. A large crack in a large sample that is closed under confining pressure acts like a collection of small cracks. Therefore, the strength of large samples and small samples under confining pressure is closer than one would predict from an examination of their microstructures at atmospheric pressure.

SUMMARY

I have tried here to describe in a concise way the major elements affecting the deformation and fracture of rock. Rock is found to be an unusual material in that it does not fall into one of the traditional classifications of elastic and plastic, describing deformation, or brittle and ductile, describing fracture. Behavior is governed to a large extent by the porosity found in nearly all rock. The opening and closing of cracks causes nonlinearity in stress–strain curves, and sliding between the faces of cracks against friction is the source of the nonelastic behavior observed under deviatoric stresses. Likewise, cracks weaken rock and determine how strength varies with applied stresses—rock is weak in tension and relatively strong in compression, and compressive strength increases dramatically with confining pressure.

REFERENCES

1. F.A. McClintock and A.S. Argon, *Mechanical Behavior of Materials*, Addison-Wesley (Reading, Mass.), 1966.

2. J.B. Walsh, The effect of cracks on Poisson's ratio of rocks, *J. Geophys. Res.*, 70(20), 5249–5258 (1965).

3. J.B. Walsh, Static deformation of rock, *J. Eng. Mech. Div., Proc. Amer. Soc. Civil Eng.*, *106* (EMS), 1005–1019 (1980).

4. M. Nishihara, Stress–strain relation of rocks, *Doshisha Kogaka Kaishi*, *8*, 32–54 (1957).

5. T.-F. Wong and W.F. Brace, Thermal expansion of rocks: Some measurements at high pressure, *Tectonophysics*, (57) 95–117, 1979.

6. T.-F. Wong, Post-failure behavior of Westerly granite at elevated temperatures, Ph.D. thesis, Massachusetts Institute of Technology, 1980, submitted to *Int. J. Rock Mech. Min. Sci.*, 1980.

7. F. Birch, Compressibility; elastic constants, *Handbook of Physical Constants* (S.P. Clark, Jr., ed.), 97–173, 1966.

8. J. Handin, Strength and ductility, *Handbook of Physical Constants* (S.P. Clark, Jr., ed.), 223–300, 1966.

9. A.A. Griffith, The phenomenon of flow and rupture in solids, *Phil. Trans. R. Soc.*, London, *A221*, 163–198, 1921.

10. A.A. Griffith, Theory of rupture, *Proc. 1st Intl. Congr. Appl. Mech.*, Delft, 55–63, 1924.

11. W.F. Brace, personal communication, 1980.

12. F.A. McClintock and J.B. Walsh, Friction on Griffith cracks in rocks under pressure, *Proc. 4th U.S. Congr. Appl. Mech.*, 1015–1021, 1962.

13. P.S.B. Colback and B.L. Wiid, The influence of moisture content in the compressive strength of rock, *Rock Mech. Sympos.* (3rd), Univ. of Toronto, 65–83, 1965.

14. W.F. Brace and R.J. Martin III, A test of the law of effective stress for crystalline rocks of low porosity, *Int. J. Rock Mech. Min. Sci.*, *5*, 415–426, 1968.

15. Z.T. Bieniawski, The effect of specimen size on the compressive strength of coal, *Int. J. Rock Mech. Min. Sci.*, *5*, 325–335, 1968.

16. H. Jahns, Measuring the strength of rock in-situ at an increasing scale [in German], *Proc. 1st Congress, Int. Soc. Rock Mech.*, Lisbon, *1*, 477–482, 1966.

17. H.R. Pratt, A.D. Black, W.S. Brown, and W.F. Brace, The effect of specimen size on the mechanical properties of unjointed diorite, *Int. J. Rock Mech. Min. Sci. 9*, 513–529, 1972.

BIBLIOGRAPHY

This short review does not do justice to these fields. Rock deformation and fracture have received considerable attention in recent years, and the literature describing experimental and theoretical studies is too voluminous to be listed here. For those who need to read further in the subject, however, the following texts will provide a good starting point:

Jaeger, J.C., *Elasticity, Fracture and Flow*, Methuen (London) and John Wiley (New York), 1956.

Jaeger, J.C. and N.G.W. Cook, *Fundamentals of Rock Mechanics*, Methuen, London, 1969.

McClintock, F.A. and A.S. Argon, *Mechanical Behavior of Materials*, Addison-Wesley, Reading, Mass., 1966.

Stagg, K.G. and O.C. Zienkiewicz, *Rock Mechanics in Engineering Practice*, John Wiley, London, 1968.

Problems in the Deterioration of Stone

ERHARD M. WINKLER

Stone decay is determined by the type of stone and by the amount and source of moisture. The carbonate rocks—limestones, dolomites, and marbles—are attacked by moisture from the surface downward; limestones tend to form a relief between dense fossil shells and a less dense matrix, with a maximum surface reduction of 0.2 mm in 10 years of exposure to 40 in. (100 cm) of precipitation annually. Crystalline marble dissolves around the grains, resulting in sanding and a rough surface relief. Secondary layers and crusts of gypsum may form by dissolution and redeposition in the presence of sulfate, a process often aided by bacterial action. The decay of silicate minerals and rocks is very slow, except for tremolite in some dolomite marbles and black mica in granites and some marbles. Black mica may form brown blotches around mica flakes, whereas tremolite decays to soft talc leaving craterlike holes in marble. Granitic rocks tend to separate into thin, even sheets parallel to the surface near ground level: Ground moisture combined with the action of salts and relief of stress from the weight of the building form this common spall, while the mineral components themselves remain unweathered.

The weathering and weathering rates of stone depend on the routes of travel and the amount of moisture, as follows: corrosive rain and drizzle on the stone's surface with a pH range of 3 to 5; rising ground moisture of variable corrosiveness, a vehicle for salt transport leading to efflorescence, subflorescence, and honeycombs; leaking indoor plumbing and gutters leading to uneven cleaning of the stone's surface and secondary deposits of calcite or gypsum, or both; and outward seepage of condensation water, leading to flaking, surface hardening, and honeycombs.

Preventing the access of moisture is the most natural but most difficult way to preserve stone.

Erhard M. Winkler *is Professor of Geology, Department of Earth Sciences, University of Notre Dame.*

This study of stone weathering was made possible by grants from the National Bureau of Standards and the National Science Foundation.

The rapid decay of stone buildings and monuments is well reflected in a pair of photos (Figure 1) of a sculpture in West Germany, taken at an interval of 60 years; the near exponential increase of the weathering rate since the beginning of industrialization is a stern warning to all of us, especially to those involved in the preservation of monuments. Test walls or similar means of monitoring susceptibility to such decay are clearly needed.

The National Bureau of Standards (NBS) was aware of the need for a stone test wall when it occupied its previous campus only a few miles from downtown Washington, D.C., in 1948. The wall (Figure 2) was moved from there in one piece—37 ft, 9 in. (11.5 m) long and 12 ft, 10 in. (4.4 m) high—to the present, more rural campus of the NBS in Gaithersburg, Maryland, in 1978. Figure 3 shows details of weathering of 4 in. (10 cm) square blocks on a section of the front (south) face. Many sandstones were most vulnerable to weathering, as shown by crumbling and scaling, whereas many limestones have developed a surface relief of about 3/4 mm between densely crystalline fossil-shell fragments and a much softer, fine-grained matrix of the same calcitic material. Most of the coping stones, the cover stones of the wall, are

FIGURE 1   A sculpture at Herten Castle near Recklinghausen, Westphalia, West Germany, carved of Baumberg sandstone in 1702. The photograph on the left was taken in 1908; the one on the right in 1969.

FIGURE 2   View of NBS stone test wall, south face. The wall is 12 ft, 10 in. high and 37 ft, 9 in. long.

Indiana limestone, which is composed primarily of fossil fragments and oolites with a calcitic bonding cement. The exposure of the north, top, and south surfaces permits the development of surface relief to be monitored with depth micrometers; these data have been correlated with wind and rain data from the nearest airport, first from Washington National Airport and now from Dulles International Airport.

The 2,400 stone samples built carefully into the NBS test wall, many in duplicate, are well protected against rising ground moisture and interior condensation. This is in contrast with stone in buildings and monuments. The origin of the moisture and its travel routes determine its effectiveness in the decay of stone, as follows:

1. Rain and drizzle, often driven against a wall by wind, are generally corrosive and acidic. The attack is primarily superficial and the pH is between 3.0 and 5.8 (rainwaters are usually charged with carbon dioxide and sulfate in urban and industrial atmospheres). The waters move in and out of the stone, dissolving ingredients within the stone and transporting them to the surface, where the waters are neutralized and redeposit the dissolved material as hard, secondary crusts. Carbonate rocks are readily attacked, with formation of a distinct surface relief.

FIGURE 3 NBS stone test wall (see Figure 2), detail section. Some sandstones show weathering, discoloration, and salt efflorescence.

Also readily attacked are porous sandstones with a calcareous grain cement; the dissolution of the grain cement may cause loss of coherence of the grain bond, while the cement itself moves outward, developing a case-hardened surface that readily scales or develops honeycombs.

2. Ground moisture travels from the ground upward by capillary action, often climbing as high as 10 meters or so. Groundwater is potentially rich in ingredients from several sources: leaching from the soil, rain running down the building into the ground, or salts used to deice streets and sidewalks. The composition of the groundwater is thus variable. The salts are carried upward to the capillary fringe, where the moisture tends to evaporate, leaving the salts behind. A "wetline" can develop, often associated with a rim of efflorescence and invisible subflorescence beneath the stone's surface. Concentration of hygroscopic salts around the wetline can lead to further attraction of moisture, especially at high relative humidities. At 90 percent relative humidity, masonry that contains 4 percent salt can retain 22 percent water.

3. Leaking plumbing, both outside and inside a building, and leaking roofs and gutters may concentrate water between ornaments and along

joints. Continuous washing, or complete prevention of it, can clean and corrode portions of the wall while other parts remain covered with soot and grime. Crusts of calcite or gypsum or both, often form beneath the washed zone. Waters of this kind vary in composition, but tend to become neutralized in contact with mortar, dust, or soluble stone.

4. Indoors, moisture from saturated air condenses on cool walls. From there the water moves toward the warm outside surface, in the process dissolving ingredients from both stone and mortar. Figure 4 shows the resulting redeposition of lime as a crust, from the mortar joint downward; the sandstone is entirely free of calcite.

Notwithstanding the foregoing description, it is difficult in most

FIGURE 4 Freiburg Cathedral, West Germany, east wall, with lime crust (calcite) from joint near window covering honeycombs in sandstone.

cases to identify the sources of moisture associated with decaying stone.

PROBLEMS OF DISSOLUTION

Carbonate rocks—such as limestones, limestone marbles, dolomites, and crystalline marbles—are readily attacked by rainwaters, especially waters charged with excessive carbon dioxide and sulfate. Limestone quarries often show channels and rills inflicted by dissolution. Examples are the Indiana limestone quarries and the Tennessee Holston limestone–marble quarries.

Dissolution is also apparent on the Georgia marble on the exterior of Chicago's Field Museum of Natural History (Figure 5). The large vertical columns framing the north and south entrances show progressive dissolution of the coarse calcite grains along the grain boundaries and along cleavage and twinning planes wherever they are exposed to the rain. Deep cracks abound along the ribs, though the foliation of the marble runs almost perpendicular to the vertical axes of the columns. No cracks can be observed on the stone that has been protected from rain. Similar cracks may be observed in columns of coarse-

FIGURE 5 Field Museum of Natural History, Chicago, south entrance. Photo shows deep weathering and vertical cracks that have developed along ribs on columns.

grained marble at other places with humid climates, such as New York City. The vertical cracks appear to be due to a combination of causes: relief of the stress of the overburden of the high columns and heavy roof; residual stress locked into the marble as a prestressed geological body (like a sleeping bag that expands when its cover is removed); and moisture–heat expansion and contraction, often combined with the action of frost. The disruptive factors are triggered by the dissolving action of acid rains; in turn, the disruption of the stone opens new channels for rain to enter and dissolve the stone. The surface reduction of the marble against unweathered hornblende shows well, although it was measured between the protecting ribs (Figure 6). The measured surface relief correlates well with the wind–rain rose.

Dissolution of soluble minerals or mineral grain cement in a porous stone is followed by the transport of the dissolved ingredients to the surface. There the solvent evaporates, leaving a thin skin or crust of calcite or silica and also soluble salts like chlorides and sulfates of sodium, calcium, and magnesium. The loss of supporting grain cement beneath the hardened surface skin causes scaling. The process is con-

FIGURE 6   Detail of weathered
protruding rib showing dislodged
calcite grains and vertical cracks at
the Field Museum of Natural History, Chicago. Scale in millimeters.

tinuous—after a scale has fallen off, another develops underneath, and so forth. The action of salt behind the hardened surface can accelerate the process of scaling. A hardened surface skin may also function as a semipermeable membrane, making way for a true osmotic pressure system. The solubility of calcite in water is well known to depend on the presence of carbon dioxide; in contrast, dissolution of silica depends on the temperature and degree of crystallinity. At 20° C crystalline quartz (silica) dissolves in pure water at only about 5 mg/l, and at 50° C about 15 mg/l; microcrystalline chalcedony is about twice as soluble as quartz, and amorphous silica (opal) has a solubility of about 100 mg/1 at 20° C and 120 mg/1 at 50° C. It should thus not be surprising that silica dissolves readily on a stone surface saturated with capillary ground moisture in the hot desert sun or on a sun-drenched masonry wall.

A headstone in the Masonic Cemetery, Fredericksburg, Virginia, shows intensive surface hardening toward the outer fringe on one side, but strong flaking with a present surface reduction of 25 mm in the center portion on the other (Figure 7). The stone is soft Aquia Creek sandstone with only a little calcite in the grain cement; the rest is mostly silica. After about 200 years of exposure, the original tool markings are still visible on the outside, while fresh stone is exposed near the center as a result of progressive scaling. In many sandstones, surface hardening may develop a honeycomb pattern in which differential hardening appears to follow a meniscus-like pattern, with crumbling occurring behind the crust and in the depressions where deepening is rapidly aided by the action of salt. Honeycombs are frequently observed on sandstones on buildings. Figure 4 shows honeycombs in a calcite-free red sandstone on the east face of the Freiburg Cathedral in West Germany. A crust of secondary calcite covers the honeycombs. The surface grain cement was introduced from the surface as calcite; it did not move to the surface to concentrate there. This case appears to be unique.

WEATHERING OF SILICATE MINERALS

Black mica, feldspars, and tremolite hornblende decay slowly, yet fast enough to enable the rate of decay to be recorded in a human generation. Black mica tends to become rusty by the oxidation of iron, which also discolors the immediate surrounding of the mineral flake. Feldspars have a distinctly glassy luster which gradually dulls as they weather to clay. Tremolite hornblende, a calcium magnesium silicate, is a common constituent of some dolomite marbles; it hydrates readily

FIGURE 7 Surface hardening on sandstone headstone at Masonic Cemetery, Fredericksburg, Va. Tool marking on upper surface is hardened and preserved while the center is flaking. Six-in. (15-cm) scale at base. Masonic Cemetery, Fredericksburg, Va.

to soft talc, leaving holes of about the original size of the mineral grains. The white tremolite is difficult to locate in white marble when fresh. In contrast, a weathered dolomite–tremolite marble is peppered with small, craterlike holes; such damage can be seen well on the south wall of the U.S. Capitol in Washington, D.C. (Figure 8).

FIGURE 8 Pockmarked dolomite marble on the southwest corner of the U.S. Capitol Building, Washington, D.C. Tremolite weathered to talc is leaving holes and causing flaking of the granite beneath.

SCALING OF GRANITES

Scaling is frequently observed on granites of medium or fine grain. Thin sheets separate readily from the stone block parallel with the outer surface, regardless of the mineral orientation of platy or prismatic components, such as mica or hornblende. The sheets are between 1 mm and 3 mm thick; their thickness is surprisingly even. Black micas and feldspars appear to be entirely fresh; they retain their original color and luster. These minerals are excellent visual indicators of the freshness of granitic rock. Scaling of granite is generally found near street level. There is strong evidence that the scaling is physical in nature. The evidence suggests that the following variables are instrumental in the formation of scales in granites:

• Expansion–contraction cycles of ground moisture entrapped in the pores of the granite.

• The action of salts introduced with groundwater and from the street. Salts in masonry attract considerable additional moisture to the stone. Although crystallization of salts at the surface often disfigures stone with efflorescence, it may also roughen the surface, which is called salt fretting. Subflorescence, the crystallization of salts beneath the surface, often leads to spalling.

• Relief of stress from the load of the building (see columns in Figures 5 and 6).

• Relief of residual or dormant stresses.

• Relief of stresses caused by machining and tooling the stone.

Figure 9 shows spalling of granite at the base of the Tweed Court House in lower Manhattan, New York City. The scales are large and thin, the minerals fresh on the insides of the flakes. Irregular thin flakes can also be seen on the granite ledge at the base of the U.S. Capitol Building, Washington, D.C. (Figure 8). The diameter of the spalled area on the Martin Luther monument in Worms, West Germany, has increased from 25 cm to about 100 cm in only 28 years of exposure, as observed by this author.

FIGURE 9 Strong flaking on granite at the base of Tweed Court House, lower Manhattan, New York.

CONCLUSION

The decay of stone in a building or monument is an extremely complex process or combination of processes that may involve several interdependent factors. Every variable should be fully understood before preservation of any kind is attempted. The first and most important step should be to identify the origin of moisture, if present, and its channels throughout the masonry. The preservative should be carefully chosen, taking into account physical and chemical compatibility with the parent stone and its potential durability under the aging process.

BIBLIOGRAPHY

Winkler, E.M., 1975, *Stone Properties, Durability in Man's Environment*, 2nd ed., Springer-Verlag, New York.

Winkler, E.M., 1977, The decay of building stones: A literature review. *Assoc. Preservation Technology Bulletin*, 9(4), 52–61.

Winkler, E.M., 1978, Stone preservation, the earth scientist's view. *Assoc. Preservation Technology Bulletin*, 10(2), 118–121.

Winkler, E.M., 1979, Role of salts in development of granitic tafoni, South Australia: A Discussion, *Jour. of Geology*, 87, 119–120.

Winkler, E.M., 1980, Historical implications in the complexity of destructive salt weathering—Cleopatra's Needle, New York. *Assoc. Preservation Technology Bulletin*, 12(2), 94–102.

Winkler, E.M., 1980, The National Bureau of Standards Stone Test Wall After 30 Years of Exposure—A Lesson in Stone Weathering. *Geological Society of America, Program with Abstracts*, 12 (7), 551.

Winkler, E.M., (in press), The effect of residual stresses in stone.

Winkler, E.M., (in press), The Stone Exposure Test Wall After 30 Years of Exposure, National Bureau of Standards.

Winkler, E.M., The weathering of Georgia marble, Chicago Field Museum of Natural History, manuscript and posters in process.

The Mechanism of Masonry Decay Through Crystallization

SEYMOUR Z. LEWIN

One of the most common and extensive sources of deterioration of stone, brick, mortar, plaster, and concrete is the consequence of crystallization phenomena that take place in pores, channels, and cracks at and near exposed surfaces. Liquid water deposits dissolved matter wherever evaporation occurs. The site of this crystallization is determined by the dynamic balance between the rate of escape of water from the surface and the rate of resupply of solution to that site. The former is a function of temperature, air humidity, and local air currents. The latter is controlled by surface tension, pore radii, viscosity, and the path length from the source of the solution to the site of the evaporation.

The detailed nature of this balance determines the form that the decay will take. If the rate of resupply of solution to the surface is sufficient to keep pace with the rate of evaporation, the solute deposits on the external surface and is characterized as an efflorescence. If the rate of migration of solution through the pores of the masonry does not bring fresh liquid to the surface as rapidly as the vapor departs, a dry zone develops just beneath the surface. Solute is then deposited within the stone at the boundary between the wet and dry regions, generating spalls, flakes, or blisters.

The site of crystal deposition can be predicted by applying the physical–chemical laws governing capillarity, viscous flow, and diffusion. These considerations disclose the quantitative relationship between the porosity of the masonry and the dimensions of the flakes, blisters, or spalls that develop, as well as the manner in which the decay progresses.

Data from controlled experiments in which salt decay is induced in laboratory specimens, together with measurements on examples of salt decay in buildings and monuments in a variety of environments, confirm the validity of these insights.

Seymour Z. Lewin *is Professor, Department of Chemistry, New York University.*

Exposed stone and other masonry materials are subject to a number of deteriorating influences, chief among which are the effects of crystallization, freezing, acidic attack, and mechanical erosion. The reality and ubiquity of the phenomenon termed "salt decay" are recognized by many of those concerned with the conservation of buildings and monuments,[1-3] but the detailed mechanism by which the crystallization of waterborne substances can break up the surface of a somewhat porous solid has not hitherto been objectively demonstrated.

When water at 0° C changes into ice, there is a volume increase of 9 percent. If liquid water is confined in a pore or crack, and this phase transformation takes place, it is evident that the resulting expansive force can damage the host solid.

It is also clear that susceptible materials can be dissolved by acidic substances generated from air pollutants (e.g., fossil-fuel combustion products), microorganisms, associated minerals (e.g., sulfides that undergo oxidation), or vegetation. Such attack can destroy surface modeling and sculpted details and weaken internal induration that binds the grains of the solid together. Similarly, the manner in which mechanical abrasion (e.g., the sandblasting effect of wind-driven dust) erodes a surface is readily visualized.

However, it is not immediately evident why the deposition of a solute from a solution into a pore or crack at the surface of a solid should damage the latter. Consider, for example, the evaporation of a sodium chloride solution at a stone surface. When, as a consequence of the escape of water vapor, the solution reaches saturation, it contains at ordinary temperatures about 26 percent solid matter by weight. Hence, a pore filled with such a solution can have only about one-quarter of its volume taken up by the residue left when evaporation is complete. Each repeated imbibition of salt solution can be expected to reduce the remaining free volume of the pore by one quarter of that value, until the pore is filled with deposited solute. But there is no analogy in this process to the expansive force that develops when water filling a pore transforms into ice or when certain types of solid phases filling a pore recrystallize into higher hydrates (as, for example, when sodium sulfate (Na_2SO_4) transforms at high humidity into $Na_2SO_4 \cdot 10 H_2O$).

Nevertheless it is a fact that the deposition of a simple, nonhydratable solid, such as sodium chloride, during the evaporation of its solution in the pores of stone and masonry, can disrupt the solid. The external manifestations of this disruption are similar to those produced by the freezing of water in the pores of the surface of the solid—scaling, flaking, and blistering and/or crumbling of the surface.

It is the purpose of this paper to investigate whether the so-called salt decay is due solely to the deposition of solute at a stone surface, to determine the conditions under which salt decay occurs, and to establish the quantitative relationships between the type of decay observed and the physical properties of the liquid and solid phases involved.

THEORY

The experimental section of this paper demonstrates that salt decay occurs only when solute is deposited within the pores of the solid—that is, a certain distance beneath the external surface (usually a fraction of a millimeter to a few millimeters). This can occur when the rate at which water departs from the surface of the solid via evaporation is equal to the rate at which fresh solution is brought to the surface via migration through the internal capillary system of the solid.

If migration of solution to the surface is faster than the rate of drying, then liquid oozes out onto the exposed surface, and solute is deposited on top of that external surface. This corresponds to the formation of visible efflorescences.[4,5] Although they may be unsightly, and usually indicate that subsurface crystallization is occurring elsewhere, they are not, per se, damaging to the stone.

If the migration of solution toward the exposed surface is very slow, then very little deposition of solutes takes place. Whatever deposition does occur is deep within the stone and does not manifest itself in the form of surface decay.

It is proposed herein that the necessary condition for surface decay is the establishment of a steady state in which the rate of diffusion of water through a thin layer of the porous solid at the surface is balanced by the rate of replenishment of water to that site from the source (reservoir) of the solution. The principle of this mechanism is depicted schematically in Figure 1.

Evaporation by Diffusion

The drying-out of solution within a pore opening at the surface occurs by diffusion of water vapor through a layer of thickness δ centimeter of the porous solid. The rate of diffusion, J grams per square centimeter per second, is expressed by Fick's first law:

$$J = D(dC/dX), \tag{1}$$

where D is the diffusion coefficient in $cm^2\ sec^{-1}$, and dC/dX is the concentration gradient across the diffusion layer.[6]

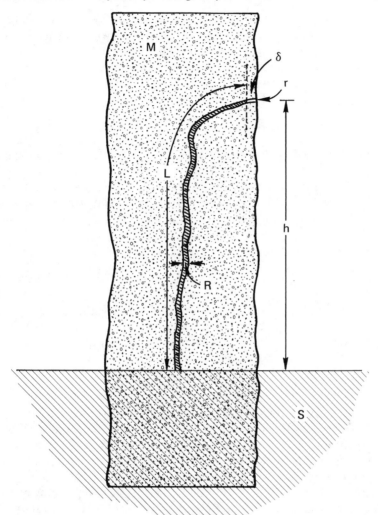

FIGURE 1 Parameters involved in the proposed mechanism for masonry deterioration due to deposition of solute from solution. The masonry, *M*, is in contact with a reservoir of solution, *S*. Solute is deposited a distance δ inside the stone at the height *h* above the reservoir. The radius of the pore opening at the site of deposition is *r*; the average radius of the channel through which solution migrates is *R*; the length of the migration path is *L*.

Because the air at the surface of exposed masonry is generally in motion, the aqueous tension at the surface of the solid tends to be constant, and the concentration gradient can be expressed in terms of the difference in vapor pressures of the water at the solution surface, P_s, and in the ambient air, P_a, divided by the diffusion-layer thickness, δ:

$$J = D(P_s - P_a)(M.W./NkT)/\delta. \tag{2}$$

In the steady state the rate of escape of water from 1 cm² of exposed surface of the porous solid is equal to the diffusion rate, J, times the fraction of open area at the solid surface, F_s. The latter is related to the porosity of the solid and is typically between 0.05 and 0.40 for natural stone and other masonry materials.[7]

Replenishment by Capillary Migration

Solution is drawn to the surface of the porous solid by capillarity. The interfacial tension, γ, at the free surface of the liquid provides the driving force that draws the liquid to the surface through the capillary network from the source. The equilibrium pressure difference at the liquid surface in a circular pore as a result of the interfacial tension is given by the Laplace equation:

$$\Delta P_s = 2 \gamma \cos \theta/r, \tag{3}$$

where θ is the contact angle of wetting of the meniscus at the walls of the pore, and r is the radius of the pore at the liquid surface.[8] If there is a distribution of pore sizes in the solid, an effective radius can be adopted that represents the weighted average of the contributions of the various pores to the resultant surface (driving) force.

This driving force draws solution to the surface to replace that which departs via evaporation. The flow of liquid through the capillary network under this driving force is governed by Poiseuille's law:

$$\frac{\Delta V}{\Delta t} = \frac{\pi R^4}{8 \eta} \cdot \frac{\Delta P}{L}, \tag{4}$$

where R is the effective radius averaged over the total length L of the capillary network through which the flow is occurring, V is the volume of solution passing through 1 cm² of pores in the time t, and η is the viscosity in poise (g sec^{-1} cm^{-1}).[9] The term $\Delta P/L$ is the total gradient of pressure from the solution reservoir to the evaporation site.

The driving force, ΔP, in an empty, uniform capillary of radius r is given by equation 3 above. As the liquid rises in the capillary, the driving force diminishes, since part of the surface pressure must provide the hydrostatic pressure to support the column of liquid of height h:

$$\Delta P_{net} = \frac{2 \gamma \cos \theta}{r} - h \rho g. \qquad (5)$$

If there were no evaporation occurring, the liquid would rise in the capillary until the surface pressure and hydrostatic pressure became equal, i.e., ΔP_{net} would be zero. When evaporation is occurring, a steady state tends to be established in which ΔP_{net} has that value which produces a Poiseuille flow just sufficient to balance the rate of escape of liquid at the evaporation site.

The driving force in the Poiseuille equation involves the effective radius r at the height h. The frictional force limiting the rate of flow involves a different parameter, R, which describes an effective radius for the entire length of capillaries through which the liquid moves to get from the source to the evaporation site.

It is shown in the experimental section that, operationally for masonry, these two parameters generally will have quite different values. The reason is as follows. Whereas the surface (driving) force involves the inverse first power of the pore radius, the viscous (opposing) force involves the capillary radius raised to the fourth power. Thus, for small capillaries (the condition for laminar flow is $R \ll 1$), the rate of Poiseuille flow falls very rapidly as the radius decreases. This has the important practical consequence that in a porous solid containing a range of pore sizes, only the upper part of the pore-size-distribution curve contributes significantly to the rate of flow of liquid to the evaporation site during the times involved in the wet-to-dry cycling of masonry in buildings and monuments. On the other hand, the surface force increases inversely as the radius decreases. Therefore, pores too small to participate significantly in the viscous flow do nevertheless make an important contribution to the net driving force drawing the liquid through the capillary network.

The Steady State

In the steady state the rate of escape of water is equal to F_s times J. The rate of replenishment is equal to the Poiseuille flow rate, $\Delta V/\Delta t$, times the fractional area of the solid that consists of contributing capillaries, F_p, times the liquid density, ρ, and weight fraction, F_w, of water in the solution. Thus:

$$F_s \, J = F_p \, \rho \, F_w \, (\Delta V/\Delta T). \tag{6}$$

Substituting equations 2, 4, and 5 into equation 6 yields:

$$\frac{F_s \, D \, (P_s - P_a) \, (M.W./NkT)}{\delta} \tag{7}$$

$$= \frac{F_P \, \rho \, F_w \, \pi \, R^4 \left(\dfrac{2 \, \gamma \, \cos \theta}{r} - h \, \rho \, g \right)}{8 \, \eta \, L}$$

and

$$\delta = \frac{8 \, F_s \, D \, \eta \, L \, (P_s - P_a) \, (M.W./Nk \, T)}{F_P \, \rho \, F_w \, \pi \, R^4 \left\{ \dfrac{2 \, \gamma \, \cos \theta}{r} - h \, \rho \, g \right\}}. \tag{8}$$

If, because of evaporation of water, solute crystallizes a distance δ beneath the exposed surface of the porous solid, and if this is the source of the deterioration of the surface, then equation 8 permits the quantitative prediction of the extent of the surface decay (i.e., the thickness of the flake, blister, or powder layer). Such prediction can be made directly and rigorously on the basis of the properties of the solid (porosity, pore-size distribution), the solution (concentration, vapor pressure, interfacial tension, viscosity, density), the solvent (diffusion coefficient, molecular weight), and the environment (temperature, relative humidity).

Deterioration from NaCl Crystallization

One of the common types of salt decay is that caused by deposition of sodium chloride in stone and brick.[3] The source of the salt may be seawater or groundwater, deicing practices, or aerosol particles.

In this case the solution just below the exposed surface tends to be a saturated sodium chloride solution; the temperature is the ambient temperature, and the solution migrating within the solid is dilute. The following values will be taken as fairly representative[10] of this system:

D = 0.22 cm^2 sec^{-1} (for water vapor in air at 1 atm and 20° C)[11]
P_s = 19 torr = 0.025 atm (for 5.3 M NaCl)

P_a = 10 torr = 0.012_5 atm (for air at 60% relative humidity)
M.W. = 18 g mol^{-1}
Nk = 82.06 cm^3 atm mol^{-1} deg^{-1}
T = 293K
ρ = 1.00 g cm^{-3}
γ = 82.0 dyne cm^{-1} (for 5.3 M NaCl)
$\cos \theta$ = 1.00
g = 980 dyne g^{-1}
$h \rho g$, the hydrostatic pressure term, will generally be negligible relative to the surface pressure term $(2 \gamma \cos \theta)/r$.

Substituting these values into equation 8 yields the following result:

$$\delta = 3.17 \times 10^{-8} \cdot L \cdot \frac{\eta}{F_w} \cdot \left\{ \frac{F_s \, r}{F_p \, R^4} \right\}, \tag{9}$$

which permits the prediction of the thickness of surface decay that will result from crystallization of sodium chloride. The prediction employs no arbitrary, empirical parameters; it is based on data on the location of the decay zone (L), the concentration of the salt in the reservoir of solution (η /F_w), and the pore characteristics of the solid $(F_s \, r/F_p \, R^4)$.

EXPERIMENTAL TEST OF THEORY

Design of the Experiment

The validity of equation 8 has been tested by a series of laboratory experiments in which the conditions during the deposition of sodium chloride at an exposed stone surface were controlled and measured. The experimental arrangement is shown schematically in Figure 2.

A rectangular sandstone column, 60 cm × 2.5 cm × 5 cm, was mounted in a glass vessel inside a Plexiglas box with its lower 5 cm immersed in a sodium chloride solution. The neck of the glass vessel was sealed with a plug of paraffin wax 1 cm thick so that liquid could not migrate up the external surface of the stone column. This served to confine all liquid migration to the internal capillary network of the stone. Salt solutions of known concentrations were fed into the glass vessel at the rate necessary to maintain the liquid there at constant level. A constant, uniform flow of air at 60 percent relative humidity

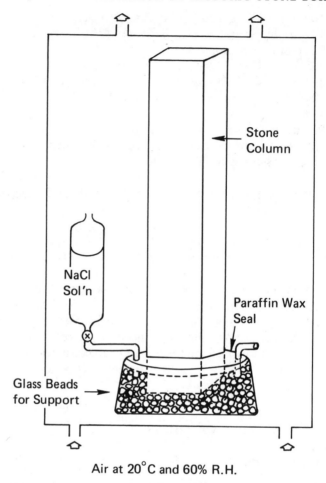

FIGURE 2 Experimental arrangement for producing salt decay. Lower end of masonry column is immersed in a salt solution, which is able to reach the exposed surface only by capillary rise through the interior of the stone. A uniform flow of air at controlled temperature and relative humidity is maintained over the exposed surface of the stone.

and 20° C was maintained over the exposed surface of the stone column.

Under these conditions a steady state was established within six to eight hours in which the solution migrated upward through the interior of the stone column to the exposed surfaces where the water evaporated, depositing the sodium chloride. The lower part of the column's surface received solution faster than it dried out, and a heavy deposit of salt formed on the stone. With increasing distance from the reservoir of solution, the thickness of the salt deposit diminished until at a certain height the surface of the stone appeared to be darker than the part above it. This indicated that the pores in the darker surface contained some liquid, but very little salt was visible on the outer surface. The typical appearance of the stone column after a two-week run with a saturated salt solution is shown in Figure 3a and with a half-saturated solution in Figure 3b.

The stone was damaged only in the region where the external deposition of salt had diminished to minute amounts—that is, where the rate of arrival of solution at the exposed surface was approximately equal to the rate of evaporation of the water, so that salt deposited within the surface rather than on top of it. The damaged surface layer of stone was held together by the subsurface crystals of sodium chloride. However, when the salt deposit was washed away, the area of decay became readily apparent as can be seen in Figure 4.

This type of experiment was conducted on the same sandstone column employing solutions of sodium chloride at three different concentrations. Before each run, the stone column was removed from the apparatus, washed free of deposited salt, soaked for two weeks in daily changes of distilled water to remove any remnants of the previously imbibed salt solution, and dried. After each run, the site and depth of the surface decay were measured. The location and thickness of the decayed zone were different for each of the different salt concentrations, as can be seen in Figure 5.

Experimental Data

Effective Poiseuille Radius

The data recorded in these controlled salt-deposition experiments relative to the surface decay are summarized in Table 1. To use these results to test the theory, it is necessary to evaluate the porosity of the stone. A type of measurement proposed here as particularly useful in this respect is the record of the rate of water imbibition as a function of the time of immersion of a test block. The data for the Longmeadow, New Hampshire, ferruginous sandstone employed in this work are shown in Figure 6.

The total porosity is estimated from (a) the volume of water imbibed after extremely long room-temperature immersion, or, equivalently, (b) that which is taken up during five hours of immersion in boiling

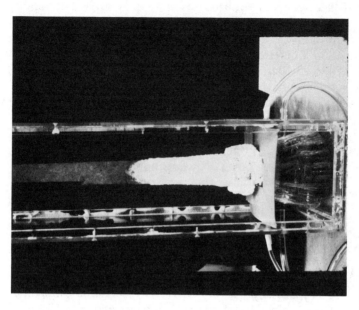

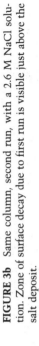

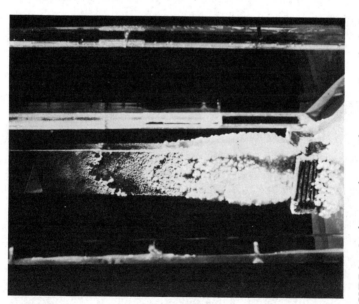

FIGURE 3a Sandstone column with encrustation of salt due to migration of solution from the interior of the stone to the surface. The triangular patch near the top shows the height to which subsurface pores appeared to be wet. Solution was 5.3 M NaCl.

FIGURE 3b Same column, second run, with a 2.6 M NaCl solution. Zone of surface decay due to first run is visible just above the salt deposit.

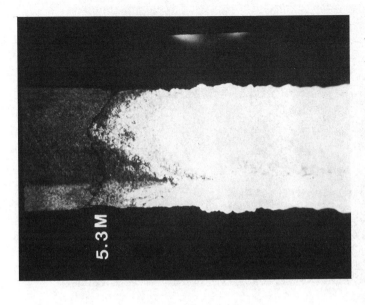

FIGURE 4b Close-up view of the stone column after the second run with a more dilute (2.6 M) NaCl solution. The decayed zone from the first run is above the zone of new external salt deposition.

FIGURE 4a Washing away the salt deposit from the first run revealed surface decay to have been produced at the place directly above the zone of heavy external salt deposit.

132

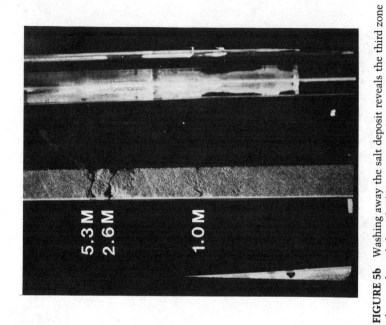

5.3 M
2.6 M

1.0 M

FIGURE 5b Washing away the salt deposit reveals the third zone of stone decay below the first two.

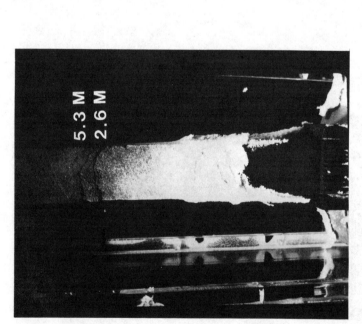

5.3 M
2.6 M

FIGURE 5a Appearance of the stone column after the third run, with a more dilute (1 M) NaCl solution as the migrating liquid. The decay zones due to the first two runs are visible above the zone of external salt deposition.

TABLE 1 Data Utilized in Test of Salt-Decay Theory

Solution	$h = L$ (cm)	F_w	η (poise)	δ calculated (mm)	δ observed (mm)
1 M NaCl	22 ± 0.5	0.98	0.0109	0.5 ± 0.3	0.4
2.6 M NaCl	27	0.95	0.0140	0.9 ± 0.5	0.8
5.3 M NaCl	30	0.89	0.0171	1.2 ± 0.8	1.0

NOTE: For the sandstone of these experiments: $F_s = 0.32$, $F_p = 0.20$, $r = 2.8 \times 10^{-5}$, $R = 1.6 \times 10^{-3}$ cm.

water followed by cooling to room temperature while still immersed, or (c) the volume taken up after the stone has been exhaustively evacuated and held under vacuum during the immersion. The estimated total porosity is 8.2 percent (i.e., 8.2 g of water is the maximum amount that can be absorbed per 100 g of the stone). This value reflects the cumulative effect of all the pores in the stone, from the largest down to those of molecular dimensions. The pore-size dispersion of this stone can be judged from the scanning electron micrographs reproduced in Figure 7.

Because, as Figure 6 demonstrates, water vapor can diffuse with approximately equal ease from the great majority of these pore openings at the surface, this datum (8.2 percent) can yield the fraction of the surface, F_s, applicable to the Fick equation. Correcting this porosity value for the relative densities of water and the stone that was utilized yields:

$$F_s = \left\{ \frac{\rho_{\text{stone}}}{\rho_{\text{water}}} \times \frac{\%_{\text{water}}}{100} \right\}^{2/3} = \{(2.20)(0.082)\}^{2/3} \tag{10}$$

$$= 0.32 \text{ cm}^2 \text{ pore space per 1 cm}^2 \text{ of stone surface.}$$

As has already been shown, only the larger pores in the distribution of pore sizes contribute significantly to the Poiseuille flow that brings solution to the exposed surface. These pores are responsible for the water imbibition that occurs during the first hour or several hours after immersion. As the data in Figure 6 show, the water imbibition was 4.0 percent by weight in the first half hour after immersion and only 4.2 percent after 3 hours of continuous immersion; a further 24 hours was required for it to increase by an additional 0.1 percent.

Thus, in the present case the porosity that contributes significantly to the steady-state Poiseuille flow may be taken as 4.2 g of water

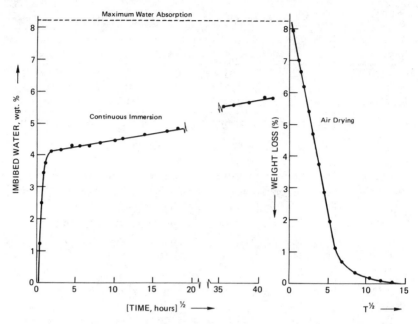

FIGURE 6 The rate of imbibition of water by a test block of New Hampshire sandstone yields information about the effective porosity for liquid flow through the internal capillary network. There is an initial rapid gain in weight due to the filling of the larger capillaries, then a very slow, diffusion-controlled process that requires many months (in fact, more than three years) to fill all the interior spaces. Immersion of the same test block in boiling water for five hours, or in room-temperature water after exhaustive evacuation, yields the weight increase shown as "maximum water absorption." The fully water-saturated test block, allowed to air-dry, loses water much more rapidly and completely than it had imbibed the water, and by a different mechanism, showing that evaporation occurs from most of the pores at the surface, whereas liquid migration occurs mainly through the larger capillaries. The test block was 7.2 × 5.5 × 15.4 cm; its dry weight was 1338.8 g; its dry density was 2.20 g/cm³.

imbibition per 100 g of stone. This value yields an effective fraction of internal cross-sectional area participating in Poiseuille transport of:

$$F_P = ((2.20)(0.042))^{2/3}$$
$$= 0.20 \text{ cm}^2 \text{ capillary area per 1 cm}^2 \text{ of} \qquad (11)$$
$$\text{stone cross section.}$$

From the comparison of total pore area with the area participating in Poiseuille flow (0.20/0.32), it follows that the upper 0.63 of the pore-size-distribution curve should be considered in evaluating the effective radius, R, to be employed in equation 8. For the present study, this part of the curve has been estimated from the scanning electron micrographs, representative examples of which are reproduced in Figure 7.

The frequencies of occurrence in 3-μm-wide pore-size intervals per 1 cm^2 of cross section in fracture surfaces have been estimated; these frequencies, multiplied by the average radius in the interval, have been raised to the fourth power; and the weighted average has been computed. The result is that the effective pore radius for Poiseuille flow in this stone is estimated as:

$$R = (1.6 \pm 0.2) \times 10^{-3} \text{ cm.} \qquad (12)$$

In this averaging technique the smallest pores—those whose dimensions were less than 0.1 μm—were not included. This does not seriously affect the validity of the resulting average, since, as has been shown, such small pores do not contribute significantly to the observed flow rate.

An alternative method of estimating the effective radius for Poiseuille flow would be to measure the rate of effusion of liquid through a plug of the stone of known dimensions under a controlled driving force and divide by the number of capillaries contributing to the total flow. This approach would also involve microscopic detection and counting of pores in cross sections of the stone and does not appear to offer any advantages of precision or convenience over the technique adopted in the present work.

Effective Laplace Radius

It remains now to estimate the effective pore radius at the stone surface that determines the surface (driving) force. This value is most reliably derived from the rate of advance of liquid through the capillary network of the stone when no evaporation is taking place, and when the hydrostatic (retarding) force is negligible. Under these conditions the Washburn equation[12] applies:

$$x^2 = \frac{\gamma \, \cos \theta \, r}{2 \, \eta} \, t, \qquad (13)$$

where x is the distance that the interfacial tension, γ, draws the liquid of viscosity, η, through a capillary of radius, r, in time, t. Figure 8

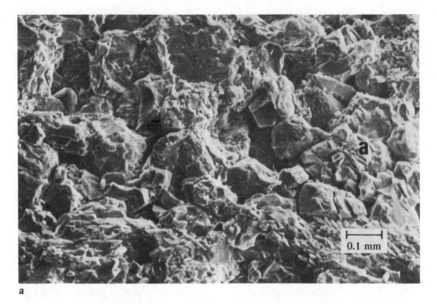

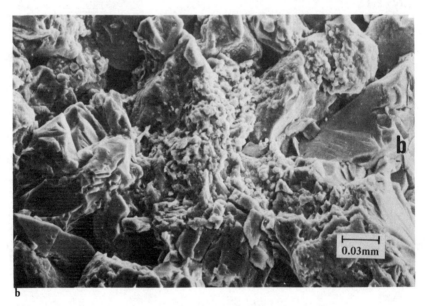

FIGURE 7 Scanning electron micrographs showing the internal pore character of the New Hampshire sandstone. Magnification employed to estimate frequencies of occurrences of pores of average radius between: **7a** 0.05 and 0.005 mm; **7b** 10.0 and 1.0 μm; **7c** 5.0 and 0.5 μm; **7d** 1.0 and 0.1 μm.

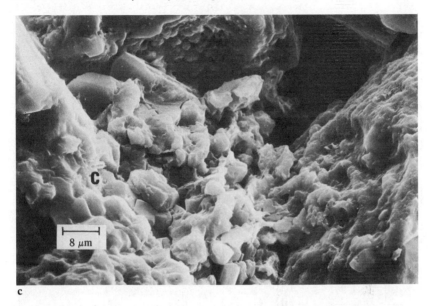

c

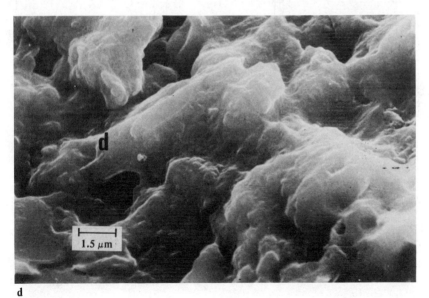

d

FIGURE 7 Continued

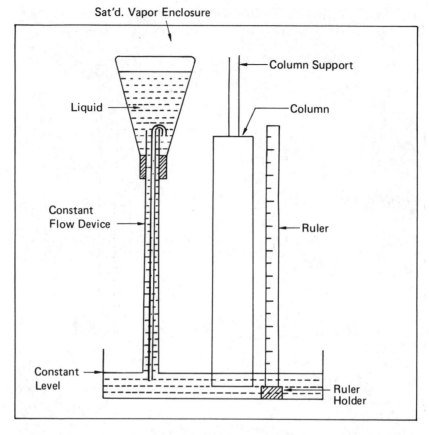

FIGURE 8 Experimental arrangement for determining the Laplace radius effective in generating the surface pressure in a masonry specimen. The height, x, to which liquid has risen in the stone column is observed visually by means of the darkening effect of wetting as a function of time, t, of contact with the bulk liquid. The inner bent tube in the constant flow device ensures that the position of contact of the liquid source with the stone column remains constant.

shows a convenient arrangement for carrying out this type of measurement,[13] and Figure 9 shows the data obtained for the sandstone under study.

The slope, m, of the graph of x versus $t^{1/2}$ is given by:

$$m = \left\{ \frac{\gamma}{2\,\eta} \right\}^{1/2} (r \cos \theta)^{1/2}, \tag{14}$$

from which an effective value of $r \cos \theta$ can be calculated. The contact angle, θ, is not known with confidence for solids such as natural stone and other masonry.[10] If $r \cos \theta$ is used in equation 8 instead of r, the derived result will be essentially equal to $\delta \cos \theta$ instead of δ. If, as is probable, the contact angle under the experimental conditions of this work is close to zero, then $\delta \cos \theta \approx \delta$. Assuming that the contact angle is close to zero, the data of Figure 9 yield the result that:

$$r = 0.28 \ \mu m = 2.8 \times 10^{-5} \ cm. \tag{15}$$

Surface Loss Calculation

For the calculation of the thickness of the surface decay according to equation 8, the interfacial tension, concentration, and vapor pressure have been taken in all cases as those of a saturated solution of sodium

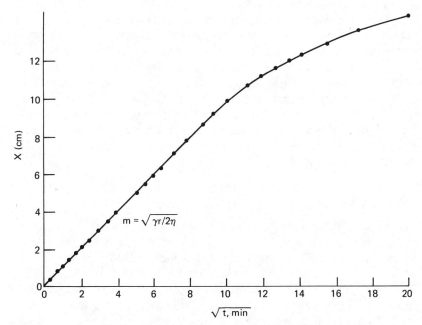

FIGURE 9 Graph of height of capillary rise, x, versus $t^{1/2}$ of contact of a New Hampshire sandstone column with a reservoir of 5.3 M NaCl solution. The slope of the straight portion of the curve yields the effective Laplace radius, $r = 0.28 \ \mu m$ (contact angle taken to be zero).

chloride, since these properties refer to the solution that is crystallizing at the stone surface and not to the solution in the interior, which may be more dilute. The internal path length, L, has been taken as equal to the vertical distance from the source of the solution to the decay zone (i.e., equal to h). That is, the tortuosity factor is taken as equal to unity for this rather porous solid.[14] The contact angle for the aqueous solutions against the polar α-quartz surfaces of this stone's pores is taken as zero degrees, and the interfacial tension is taken as that of a saturated NaCl–glass interface.

The remaining relevant data are collected in Table 1, which also compares the values predicted by equation 8 with those observed experimentally. The largest source of uncertainty in this test of the theory is the estimation of the effective Poiseuille radius, R. It will be noted that the experimental results agree very satisfactorily with the predicted values, within the precision of the data.

In these experiments the order of magnitude of the thickness of the deteriorated surface layer resulting from these salt solutions in this particular sandstone proves to be between a fraction of a millimeter and one or two millimeters. Our observations, and those of others, in studies of the decay of exposed stone and masonry in buildings and monuments have disclosed that the natural decay of many other sandstones and other types of natural stone and masonry results in surface losses of the same order of magnitude. That is, when initial salt decay is indicated—i.e., when a single layer of stone has been lifted up in the form of a blister, spall, or flake—the thickness of the deterioration is in the vicinity of a millimeter. Thus, it appears that the parameters characteristic of the present experimental setup are similar to those commonly encountered in practical instances of crystallization-induced deterioration of masonry.

CONCLUSIONS

The present work demonstrates that the mechanism of the so-called salt decay of exposed stone and masonry consists in the deposition of solutes from solution within the pores of the solid close to the surface. This is characteristically manifested in the form of a thin layer of the surface that lifts up in the form of a blister, peels outward as a spall, flakes off, or powders away. The initial thickness of this surface decay is of the order of a millimeter. When this thickness of surface has separated, the decay process may be initiated again in the underlying, still sound stone, resulting in a second such decay layer under the first. The processes can then proceed again beneath these layers, and so on.

In some cases, many successive layers of decay can be recognized, all of them similar in character, with thicknesses from a fraction of a millimeter to 1 or 2mm. An example of the occurrence of blisters 1 mm thick on exposed granite is shown in Figure 10; examples of the multiplication of decay layers, progressing from the outer surface of the exposed stone toward the interior, are shown in Figure 11.

The necessary condition for the occurrence of this type of decay is the development of a steady state at the exposed surface, wherein

FIGURE 10 Granite surface stone in the lower course of a New York City landmark building has been subjected to the action of salt used in de-icing the adjacent street. The surface has lifted up in numerous places, forming blisters with a layer thickness that ranges from 0.5 to 1.5 mm.

FIGURE 11a Cross section of the surface layers of a salt-decayed sandstone.

FIGURE 11b A sandstone sculpture in the sculpture garden of the Brooklyn Museum, New York City, showing the development of multiple layers of surface decay, resulting from successive salt-decay processes proceeding from the outside inward.

the rate of evaporation of water via diffusion through a layer of the porous solid is balanced by the viscous flow of solution from the reservoir to that site through the internal capillary network. The quantitative relationship between the thickness of surface deterioration and the characteristics of the liquid and solid media are derivable from classical physical chemistry via the Fick and Poiseuille laws. The parameters needed to describe the porous nature of the solid are obtainable from water imbibition, capillary rise, and pore-size-distribution measurements. Laboratory experiments conducted under controlled conditions yield results that are in good agreement with the predictions of this theory.

These considerations, and the related experimental observations, establish beyond reasonable doubt that the deposition from solution of a simple, nonhydrated salt, such as sodium chloride, in the pores at the surface of a stone generates pressures sufficient to break down the induration. We are convinced of the reality of the phenomenon and can now account for it in detail and predict where and under what conditions it will occur. We do not yet understand how the requisite disruptive pressures can be developed in the pores of the solid. The fundamental question that remains to be addressed is: How do crystals that have grown from a solution until they fill the volume of a pore continue to grow at the areas of direct contact between crystal and pore wall?

REFERENCES AND NOTES

1. S.Z. Lewin and A.E. Charola, Scanning Electron Microscopy in the Diagnosis of "Diseased" Stone, *Scanning Electron Microscopy, 1978*, vol. I, pp. 695–703, SEM, AMF O'Hare, Ill.

2. A.E. Charola and S.Z. Lewin, Examples of Stone Decay Due to Salt Efflorescence, Third International Congress on the Deterioration and Preservation of Stones, Venice, 24 October 1979, in press.

3. S.Z. Lewin and A.E. Charola, The Physical Chemistry of Deteriorated Brick and Its Impregnation Technique, *Congress for the Brick of Venice, 22 October 1979, Venice, Proceedings*, pp. 189–214, University of Venice, Italy.

4. S.Z. Lewin and A.E. Charola, Aspects of Crystal Growth and Recrystallization Mechanisms as Revealed by Scanning Electron Microscopy, *Scanning Electron Microscopy, 1980*, vol. I, pp. 551–558, SEM, AMF O'Hare, Ill.

5. A.E. Charola and S.Z. Lewin, Efflorescences on Building Stones; SEM in the Characterization and Elucidation of the Mechanisms of Formation, *Scanning Electron Microscopy, 1979*, vol. I, pp. 379–387, SEM, AMF O'Hare, Ill.

6. W. Jost, *Diffusion in Solids, Liquids, Gases*, Academic Press, N.Y., 1952, 8ff.

7. A.E. Scheidegger, *The Physics of Flow Through Porous Media*, Univ. of Toronto Press, 1960, p. 13.

8. F.A.L. Dullien and V.K. Batra, Determination of the Structure of Porous Media, in *Flow Through Porous Media*, American Chemical Society, Washington, D.C., 1970, 17ff.

9. T.L. Poiseuille, *Compt. Rend., 11*, 961 (1840); *12*, 112 (1841); *15*, 1167 (1842).

10. There is some uncertainty concerning the appropriate value of the contact angle, θ, for this system. The data of D.D. Eley and D.C. Pepper, *Trans. Faraday Soc., 42*, 697–702 (1946), suggest that $\theta = 0$ degrees for water in plugs of powdered Pyrex glass. The data of B.V. Deryagin, M.K. Melnikova, and V.I. Krylova, *Colloid J. USSR, 14*, 459 (1952), suggest $\theta = 60$ to 70 degrees for water spreading through a packing of quartz sand.

However, advancing contact angles tend to be different from receding angles, which tend to be zero; cf. W. Rose and R.W. Heins, *J. Colloid Sci.*, *17*, 39 (1962).

11. A. Winkelmann, *Wied. Ann.*, *22*, 1, 152 (1884); *23*, 203 (1884); *26*, 105 (1885); *33*, 445 (1888); *36*, 92 (1889).

12. E.W. Washburn, *Phys. Rev.*, *17*, 273–83 (1921); see also V.G. Levich, *Physicochemical Hydrodynamics*, Prentice-Hall, Englewood Cliffs, N.J., 1962, pp. 382–383.

13. J.N. Chan, Gypsum Plaster as a Prototype in the Study of the Physical Chemistry of Solid Porous Media, Ph.D. thesis, New York University, October 1980.

14. The tortuosity factor, T, is defined empirically as the correction factor needed to make the calculations for certain theoretical models of pore structure agree with experimental data. F.A.L. Dullien calculates, based on his particular model of porosity, that the tortuosity for sandstone would be in the range of 1.5 to 1.7; see *AIChE J. 21*, 299 (1975). However, he points out that "using different models of pore structure, widely different values may be obtained for T, some of which completely lack any physical meaning." (See *Porous Media Fluid Transport and Pore Structure*, Academic Press, N.Y., 1979, p. 227). It may be noted that if L is significantly greater than h (i.e., $T > 1$), the calculated value of δ will be too large by that tortuosity factor. This effect is in the opposite direction and would tend to offset any overestimation of δ due to the contact-angle factor (cf. ref. 10).

Characterization of Bricks and Their Resistance to Deterioration Mechanisms

GILBERT C. ROBINSON

Brick and mortar are building materials of excellent durability, but they are subject to deterioration processes that can reduce their effectiveness. The rate of deterioration is a function of composition, pore structure, manufacturing procedure, structural design, and cleaning procedure. Deterioration of masonry results from several mechanisms, including freezing and thawing, salt crystallization, chemical attack by water and other substances, moisture expansion, other internal expansive reactions, and mismatch in dimensional characteristics of wall components.

The susceptibility of brick to each mechanism is determined by pore structure and composition. The glassy and amorphous phases are key items of composition. Illustrations are presented of the significance of glass-phase composition. The manufacturing procedure, structural design, and cleaning procedure can exert additional influence on deterioration mechanisms and rates. The danger of waterproof coatings is presented.

Archeologists study ancient cultures by examination of brick and other fired clay artifacts that have persisted hundreds and thousands of years. Brick has an excellent record as a durable, versatile, and attractive building material. This is attested to by such buildings as Monticello, the Old North Church in Boston, and the Wren building at the College of William and Mary. Nevertheless, bricks do change with age, and

Gilbert C. Robinson *is Professor and Head, Department of Ceramic Engineering, Clemson University.*

the ravages of time produce varying results. In many cases the aging produces minor changes such as small shifts in color or accumulation of surface dirt. More severe damage may occur in other instances, with crumbling and disintegration or gross cracking of masonry units. The extent of change is determined by the severity of environmental exposure, the structural design, and the properties of the bricks.

The properties of brick are determined by manufacturing procedures, and these have changed over time. It is important to consider this influence when contemplating restoration procedures or the likelihood of success in restoration.

MANUFACTURING METHODS

Raw Materials

Raw materials have changed as the brick industry has evolved. The earliest bricks were made from clay similar to that used in making earthenware utensils. Clay is a rock or an integral part of the earth's crust that is composed of clay minerals and accessory minerals such as quartz, feldspar, and calcite. The clay mineral portion is made up of particles essentially smaller than 2 μm; the accessory minerals may range from 2 mm down to 2 μm. This clay was the major raw material for brick made in the first 100 years of this country's history. (Later, there was a marked switch to shales and similar rocks. These consolidated materials offered certain advantages in manufacture.) The use of fireclays was a later raw materials development. These are coal-measure clays that produce light, fired colors of ivory, yellow, and gray; Pennsylvania and Ohio were major producing areas. The most recent addition to the raw materials mix has been kaolin and sericite schist, used by brick manufacturers in the Southeast.

Shaping

Bricks were hand-molded during the early years of this country. Sufficient water was added to the clay to produce a soft plastic mix, and the material was thrown into rectangular molds. Later (1793–1819) machine-assisted equipment was developed to simulate the hand-molding, and the equipment would drop, throw, or vibrate the wet clay into the mold cavity.[1] This type of molding began to decline with the development of extrusion equipment, and today only a small percentage of building bricks are made by this method.

Extrusion equipment was developed about 100 years ago.[1] In this equipment an auger propels plastic clay through a die opening that forms the cross-sectional outline of the brick. Such machinery permits shaping brick from clay of much stiffer consistency, and recently developed, high-horsepower machines produce an extruded column of high strength. It is possible to stack the bricks one on top of the other for travel through the dryer and the kiln, rather than drying them on pallets as required by the molding process.

There has also been limited production of pressed brick. In this process, the clay is pressed into the mold cavity. The clay may be of a plastic consistency or may contain less water than required to develop plasticity.

Firing

The firing process is the key step in manufacturing brick. During firing, the porosity of the product is reduced, and a bond is developed between the particles by partial fusion and/or sintering of amorphous constituents. Firing is responsible for the development of strength in the brick. It is also responsible for color and resistance to disintegration by rainwater, freezing and thawing, or other disruptive forces. Thus the quality and extent of firing determine the major characteristics of the brick.

Firing equipment was crude in the early history of this country. Bricks were normally stacked together in an appropriate configuration in a field. The dry bricks formed the walls and firing eyes of the kiln as well as its load. Wood was used to fire these field kilns, resulting in large variations in temperature from one section of the kiln to another and a corresponding variation in the properties of the bricks produced.

Various types of kilns evolved later: Permanent walls of fired brick were constructed for field kilns; periodic kilns were developed; and the round, down-draft kiln became popular. This unit could produce much more uniform bricks than were obtained from field kilns. The tunnel kiln started a rapid ascendancy about 1940 and by the 1950s had largely replaced periodic kilns. The tunnel kiln produced exceptionally uniform properties and allowed much shorter firing cycles—perhaps 30 hours, compared with 5 days in the periodic kiln.

Coloration

During the early period of production, the fired color of brick was determined by the clay raw material. A variety of colors were obtained by varying firing temperature and kiln atmosphere (flashing). Later, and particularly during the past 20 years, there has been a growing production of brick colored by coatings of different composition than the body of the brick. The coatings range from a sprinkling of natural or colored sand through mineral slurries to impervious glazes.

INFLUENCE OF MANUFACTURING PROCEDURES ON PROPERTIES OF BRICK

Shaping Method

Bricks prepared by the molding process show higher porosity and larger pores than do extruded brick (Figure 1). Larger pores seem to enhance durability. The acceptable porosity is higher for molded brick than for extruded brick as a result of this difference in pore structure.

Firing Method

A large proportion of the charge in field kilns would turn out to be soft bricks of high porosity. The softer units could be scratched with a steel knife blade and would exhibit orange or salmon colors. The

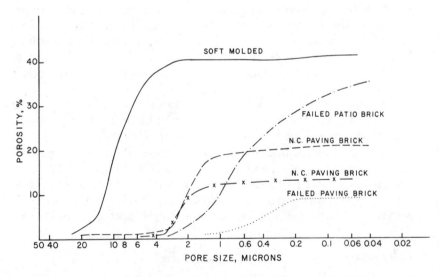

FIGURE 1   The pore-size distribution of a soft molded brick of good durability compared to two extruded bricks that failed in service and two durable paving bricks.

bricks in the higher-temperature zone of the kiln would be harder than steel, would have a good ring when struck together, and would be a darker brick-red or even brown or black. The variation in properties posed no problem at the time; the softer units were used in interior walls and the harder units for facing. Any reconstruction that exposes these interior bricks of former periods to the outdoor environment is poor practice; they probably lack the properties needed to survive in freezing and thawing environments.

The extent of firing can be estimated by examining selected properties of brick. Strength and hardness increase, absorption decreases, and color changes with increasing temperature or time at temperature. Properties from the surface of a brick to its center become more uniform with increasing time at temperature. An oxygen-deficient kiln atmosphere produces equivalent properties at lower temperature than does a kiln with an oxidizing atmosphere.

The pattern of properties at various firing temperatures is distinctive for each raw material. Table 1 shows patterns for selected materials.

AGING MECHANISMS

Several processes operate over time to alter the properties of structures made of brick. The extent of the alteration depends on the ability of

TABLE 1 Fired Properties and Durability of South Carolina Kaolin and North Carolina Shale Brick Fired to Selected Maturing Temperatures

Sample	Firing Temp. (°F)	Absorption (%)		Saturation Coefficient	Initial Rate of Absorption	Apparent Density (g/cc)	Durability* (Cycles to Failure)
		Room	Boiling				
SC kaolin	1900	15.1	16.2	0.93	41	—	4
	2000	12.9	14.2	0.90	36	2.84	5
	2010	11.3	12.8	0.88	31	2.50	5
	2050	7.9	10.0	0.79	14	2.26	10
NC shale	1900	14.2	16.0	0.89	59	2.73	5
	1950	11.0	13.0	0.85	45	2.71	5
	2000	9.2	11.4	0.81	46	2.68	10
	2050	7.6	9.8	0.77	32	2.66	10
	2150	4.3	6.6	0.65	11	2.58	+10

*Durability was evaluated by a salt (sodium sulfate) crystallization test.[2] The number of cycles to cause failure are listed. The +10 means it withstood 10 cycles without failure.

the brick to withstand weather, chemical attack, and the effects of improper structural design or execution.

Water Penetration and Bond with Mortar

The ability to resist water penetration is a major determinant of the stain resistance and durability of brick structures. Dry walls will not effloresce, break up from freezing and thawing, or be subject to chemical attack.

The two primary sources of water penetration are inadequate flashing and roofing and a poor bond between mortar and brick. The permeability of the mortar or brick represents a minor but possible source of water penetration. It is essential to correct the sources of water penetration if other restorative procedures are to succeed.

The performance of a masonry wall depends on the properties of the mortar as well as on the properties of the brick and the compatibility of these two constituents. The properties of mortar important to bond development are water retentivity, workability, and tensile strength. Interfacial tension determines bond strength developed by the interaction of brick and mortar. However, these properties are overshadowed by the practice of the mason and the weather, the predominant determinants of bond development.

Mortars have changed with the period of history. In some early structures they were made of clay and sand, with perhaps a surfacing of lime mortar. Sometimes natural substances associated with the clay (such as gelatinous silica, alkali silicates, and marl) provide substantial resistance to weathering. More often the clay bond is susceptible to disintegration by water attack (slaking). Erosion or removal of the high-lime surface exposes the interior clay mortar to disruptive water attack.

Lime sand mortars have been used since early times and once were the predominant bonding medium. Later, portland cement was added to lime mortars; today, prepared proprietary mixtures called masonry cements are commonplace. This trend in composition has produced mortars of higher compressive strength and more rapid strength development, but lower water retentivity and workability. As a consequence, the development of a good bond between brick and mortar has become more difficult and more demanding on the properties of the brick. The use of a modern mortar with an old brick may be poor practice. Present day mortars lack the flexibility in structure and self-healing characteristics of early lime mortars. Another recent trend has been the use of air-entraining agents to improve the workability of mortar. These agents increase the porosity and permeability of mortar and reduce its compressive strength.

The properties of brick significant to bond development are the surface pore structure, the initial rate of adsorption, and the tensile strength. Mechanical gripping is the major bonding mechanism between brick and mortar. The mortar paste enters the exposed surface pores and sets. The surface pores should be large enough to allow penetration and should provide undercuts, such as a spherical opening with a narrow neck, to assist adhesion. Cut surfaces on extruded brick can produce shaggy overhang projections that assist adhesion. Die-slickened surfaces are less conducive to bond development.

Other bonding mechanisms exist. Glassy brick with no surface pores will develop a bond strength of 10 psi. The same brick produced to give open surface pores will develop a bond strength of 100 psi.[3]

The initial rate of absorption (IRA) of brick is determined by the standard ASTM test C67,[4] which indicates the speed with which the brick withdraws water from the mortar. This characteristic has a pronounced influence on bond strength. The capillarity of the brick should be sufficient to pull mortar paste into the pores of the brick but not high enough to dewater the mortar before it penetrates the pores. The IRA is determined by immersing one face of a dry brick to a depth of 3.18 mm in water for 1 minute. The water absorbed is expressed as grams per brick (per 194 cm^2 of immersed area).

It has been found that good bond strength will develop with IRAs of between 10 and 40 g.[5] Figure 2 shows the relationship between bond strength and IRA. Mortars with higher water retentivity show good bond strength at higher values of IRA, while lower water retentivity depresses the relationship. Prewetting a brick to reduce its capillary suction may lead to a bond strength characteristic of the lower IRA thus induced.

Laying of frozen or saturated bricks will result in the poor or non-existent bond characteristic of brick with an IRA of zero. Weather conditions that promote drying may interfere with bonding and hydration of the mortar's constituents. Freezing weather may disrupt the mortar as well as interfere with bond development.

Exposure to Freezing and Thawing Conditions

The saturation of brick or other building units with water followed by freezing can produce disruptive forces that will destroy the units. This is a consequence of the 9 percent volume expansion of water as it changes to ice and the hydraulic pressures generated ahead of the ice.

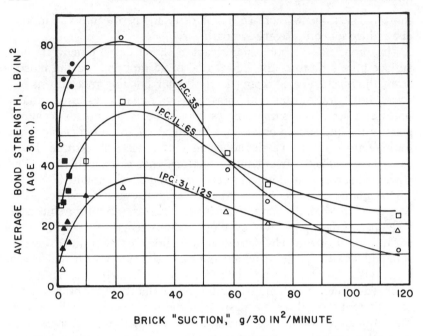

FIGURE 2   Bond strength versus IRA for mortar of different water-retaining capacity.[5]

Any structural configuration that entraps water in a void or impedes the flow of water in advance of a moving ice front will cause damage. The failure normally appears as a crumbling or disintegration or a delamination of the unit (see Figure 3).

Most brick can withstand freezing and thawing without damage, but some units with inadequate properties will fail. Certain properties are keys to this type of damage. One is porosity. The porosity of brick ordinarily is evaluated by determining the weight percentage of water it absorbs. Water absorptions in excess of 12 percent begin to suggest deterioration by freeze–thaw mechanisms. The greater the absorption, the greater the likelihood of damage; usually, absorptions above 15 percent are unacceptable for extruded brick.[6] Pressed and molded brick will give acceptable performance with higher absorptions, perhaps 14 to 17 percent.

Porosity, as indicated by absorption, is not the sole determinant of durability. Pore structure also is significant. Early attempts to evaluate pore structure used saturation coefficient as a measure of the proportions of large pores and small pores. Present efforts are directed toward determining pore-size distribution, but the saturation coefficient remains a useful if imperfect index of the durability of brick.[6,7]

FIGURE 3 The appearance of brick disintegrated (upper) and cracked in lamination planes (lower) by exposure to freezing and thawing.

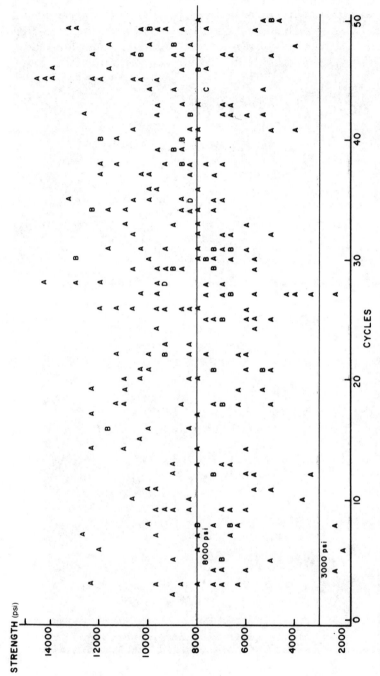

FIGURE 4 Strength versus number of freeze-thaw cycles to produce failure.[6]

The saturation coefficient is the ratio of water absorbed during 24 hours of immersion in room-temperature water to that absorbed during 5 hours of immersion in boiling water. The room-temperature absorption is supposed to indicate the amount of water a brick will pick up from rainfall, while the added absorption from boiling indicates the quantity of pores that can be penetrated only under abnormal pressure. These pores, or voids, should serve to relieve the hydraulic pressures in advance of a freezing front passing through the brick.

Saturation coefficients below 0.75 give good assurance of durability. However, there are examples of brick with excellent durability having values above 0.85. In these instances, other factors of manufacturing procedure are predominant in determining durability.

Extruded bricks exhibit a laminar structure. This structure may be made up of thin, flaky elements, or it may show a large, spiraling crack within the unit. Attempts have been made to relate lamination to lack of durability, but the relationship is questionable. A brick with unfavorable pore structure will delaminate under repeated cycles of freezing and thawing; however, the cause of failure is the inadequate pore structure and not the laminar structure. It is interesting to observe that materials that do not have an extruded laminar structure will still fail by delamination. Soft molded brick and granite are two illustrations of this mode of failure.

Strength seems logically related to freeze–thaw resistance. Nevertheless, attempts to correlate compressive strengths with durability have been unsuccessful. Figure 4 shows freeze–thaw resistance versus strength for commercially manufactured units. It can be seen that there is no correlation. The lack of correlation is believed to result from faults in the method of measuring compressive strength and perhaps from the use of the wrong type of strength determination.

There has been little investigation of the permeability of bricks. However, it has been suggested that the speed with which water penetrates a brick bears some relation to durability. Experiments have shown some relation between IRA and durability (Table 1), and the IRA test has proven useful for predicting durability. Other work is being conducted to obtain more information on the role of permeability in freeze–thaw resistance.

The properties of brick are usually determined on a whole or a half brick, and the values reported are really averages for the entire brick. Sampling at different locations from the surface to the interior of the brick will show different properties, including different pore structures. These differences have become particularly significant with the fast-firing schedules of modern tunnel kilns. High production rates allow insufficient time for heat to penetrate to the interior of the unit, which can cause property differentials. The property differentials will be greater for a solid unit than for a cored unit because the heat must penetrate more material.

Salt Crystallization

Saturation of a building brick with salt solution can produce failure similar to that resulting from freezing and thawing. The salt will crystallize within the pores of the unit and with repeated wetting and temperature cycling may produce an expansive force that will disintegrate the unit. This phenomenon can be observed in structures subjected to saltwater spray. Figure 5 shows solar screen tiles that have been disintegrated by this mechanism in Miami Beach. It is interesting that the mortar withstood the assault better than the bricks. Evaluation of the bricks showed them to have been underfired. They had an absorption of 18 percent and a saturation coefficient of 0.95 and were soft and easily scratched. They contained 14 percent dissolved salt. This type of brick, however, might be satisfactory for back-up or interior locations or even exposed locations in nonfreezing environments that are free of soluble salts.

There are other sources of salts within masonry structures. Brickwork around flower beds or gardens may collect soluble salts from

FIGURE 5 Salt disintegration of solar screen tile in Miami Beach, Florida.

fertilizers. The use of calcium chloride to prevent freezing of mortars will introduce a large quantity of soluble salts that may concentrate in one part of the masonry structure. Cleaning agents dissolve cement and other substances to produce soluble salts; injudicious use of these agents can contribute to increased salts within the masonry structure.

Chemical Attack

Properly fired bricks have unusual resistance to attack by chemical agents. They can form containers for hydrochloric acid and are subject to disruptive attack only by hydrofluoric acid. However, bricks that are inadequately fired are subject to attack. Even water will react slightly with the glassy constituents of a highly porous brick. Such a reaction will produce some reduction in the strength of the unit and contribute to moisture expansion. This process is usually not continuous; instead, the attack proceeds to a certain limit and then stops without further disintegration.

Acidic solutions will accelerate such an attack. Thus, the formation of sulfuric acid solutions from atmospheric constituents will increase the dissolution of the brick approximately 10-fold. Even in this circumstance, however, the maximum dissolution of the brick is usually about 1 percent. The resulting salts are more damaging to the appearance of the building than to its strength. These salts are a source of staining and efflorescence on the masonry structure. The sulfur trioxide required for this attack can come from pollutants in the atmosphere, but can also come from the masonry itself. The production of bricks in a high-sulfur atmosphere may yield residual sulfates that will become acidic in the presence of water. The cement constituents of mortars may be relatively high in sulfates and alkalies, and solutions of these constituents may accelerate the attack on the brick.

Moisture Expansion

Most porous materials will exhibit some moisture expansion. The moisture expansion of brick is usually slight, about 0.04 percent. In a few instances, underfiring may give higher expansions (perhaps 0.1 percent). Particles of lime or gypsum within the clay raw material can cause still higher moisture expansions, at times sufficient to disintegrate masonry units or crack and destroy walls. Also, reactions between brick and alkali solutions from the cement or other sources may result in large expansions.

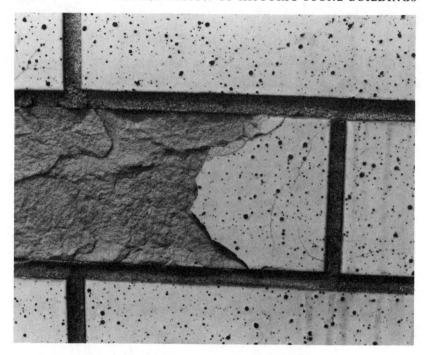

FIGURE 6 Surface spalling resulting from application of water-impermeable coating to a wet wall.

Water-Impermeable Coatings

Water-impermeable coatings can be a factor leading to damage of masonry units. Water may enter the interior of the wall and dissolve soluble salts, which then migrate toward the barrier and crystallize behind it. The exertion of a disruptive force will pop the waterproofed face away from the masonry unit. Thus, it is particularly dangerous to apply a waterproof coating over an entire masonry wall. This practice may result in no damage, however, if the interior of the wall is dry and no water can enter the interior from other sources, such as the roof. If water does enter, salt crystallization will cause severe damage in the wall (see Figure 6).

Structural Design

A common source of damage to masonry walls, in those few instances where damage occurs, is inadequate design of the structure. Inadequate flashing at the roof line or other systems that allow water to penetrate the interior of the wall can be quite damaging. Any restoration procedure should emphasize proper construction techniques to prevent leakage of water from the roof or other structures to the interior of walls. Another source of damage is inadequate expansion joints that permit unrestrained movement of the masonry as a result of temperature cycling.

Mismatch of Materials

It should be recognized that steel, concrete, and brick masonry show different dimensional behavior with fluctuating temperatures and from shrinkage. These differences can lead to cracking of a masonry wall. Figure 7 shows brick masonry fitted tightly around a metal fixture for holding electric lights; the difference in dimensional behavior between the metal and the brick has caused cracking within the brickwork. Restoration procedures should avoid this type of mismatch and correct any existing mismatches.

Cleaning Procedures

Cleaning brick masonry walls generally can be accomplished with the use of water alone or in combination with chemical reagents or sometimes by sandblasting.

FIGURE 7 Cracking of masonry from mismatch in dimensional behavior of metal light box and brickwork.

It should be recognized that chemical cleaning or sandblasting has the potential for causing serious damage to the masonry. Cleaning actions depend on solution or abrasion of the offending substance, and these actions are available also for attack of the masonry. Furthermore, chemical cleaners produce soluble salts that may penetrate the masonry and cause development of new discoloration or contribute to new deterioration. The injudicious selection and/or mixing of various cleaning reagents can cause worse staining than that originally present. The extent of damage is influenced by the concentration and quantity of cleaning agent and the duration of the cleaning procedure. The susceptibility of the masonry to attack will depend on its composition and manufacturing history. Thus, lime in mortar is more susceptible to solution than Portland cement, and underfired bricks are more readily attacked by abrasion and reaction than properly fired units.

The many variables that influence cleaning make it important to follow recommended cleaning practices[7,8] and to pretest the selected procedure in a small inconspicuous place on the masonry wall. Observations should be made of the cleaning effectiveness and of any unacceptable discoloration or damage to the masonry.

RESTORATION PROCEDURES

Restoration procedures should be selected after examination of the masonry structure. Visual examination may be sufficient to indicate staining problems: the mode of failure, if any, of the brick; sources of water penetration, such as cracks between mortar and brick; unfilled joints; wall cracks; and inadequate flashing or roofing.

In other instances, sampling and testing of the wall may be required. The scratch–hardness test remains a simple but effective means of establishing the firing history and soundness of brick. Bricks or pieces of brick can be removed from a wall and their absorptions and saturation coefficients determined to provide additional clues to durability.

Determining the possibility of increasing the structural load on a building is a complex problem. There is probably no sure way to determine the load-carrying potential of a structure short of loading it to destruction. Strength testing of small sections of a wall has limited value because of the large variation in properties of brick produced in earlier times. Table 2 shows the strengths of individual bricks removed from the capitol of Florida. The strength varies from good to nonexistent. However, even if all the bricks have the highest strength, a wall might be weak because of gaps in the mortar joint in some locations. The variabilities in properties and construction practices make

TABLE 2 Strength and Absorption of Brick from a Candidate Building for Restoration

Year of Construction	Unrestrained Compressive Strength (psi)	Absorption (%)
1902	1168	19
1902	1900	22
1845	494	26
1845	0	—

it difficult to predict the maximum permissible loading for a structure; one must allow large safety factors and use judgment based on experience in such cases.

To prevent further deterioration of a damaged structure, the cause of deterioration must be removed. For example, structural stress caused by mismatch of dimensional movements among different materials must be eliminated. Walls should be dry before waterproofing agents are added.

The use of coatings to strengthen units has been studied by numerous authors. Lewin and Charola suggest alkoxysilanes as the most promising coatings.[10] Lal Gauri discusses the use of fluorocarbons and a series of polymer solvent mixtures with increasing concentration of polymer.[11] Drisko discusses different waterproofing materials and gives recommended coatings for different situations.[12]

Cleaning procedures should be selected to suit the stains and materials involved. Any procedure should be tested at a small, inconspicuous place before it is applied to the entire structure.

REFERENCES

1. P.E. Jeffers, The Building of America, *Brick and Clay Record*, 169(1):19–30 (1976).
2. G.S. Robinson, An Accelerated Test Method for Predicting the Durability of Brick, MS thesis, Clemson University, Clemson, S.C. (1976).
3. H.D. Martin, Adhesion Mechanisms in Masonry Mortars, MS thesis, Clemson University, Clemson, S.C. (1965).
4. Standard Methods of Testing Brick and Structural Clay Tile, ASTM Designation C67–78, *1980 ASTM Annual Book of Standards*, Part 16, pp. 45–54 (American Society for Testing and Materials: Philadelphia, Pa., 1980).
5. L.A. Palmer and D.A. Parsons, A Study of the Properties of Mortars and Bricks and Their Relation to Bond, *Natl. Bur. Stand. J. Res.* 12 (1965).
6. G.C. Robinson, J.R. Holman, and J.F. Edwards, Relation Between Physical Prop-

erties and Durability of Commercially Marketed Brick, *Am. Ceram. Soc. Bull.* 56(12): 1071–1076 (1977).

7. B. Butterworth, Frost Resistance of Bricks and Tiles: A Review, *J. Br. Ceram. Soc.* 1(2): 203–223 (1964).

8. *Cleaning Clay Products Masonry, Technical Notes on Brick and Tile Construction*, 20. Brick Institute of America, McLean, Va. (1964).

9. *Good Practice for Cleaning New Brickwork*. Brick Association of North Carolina, Greensboro, N.C. (no date).

10. S.Z. Lewin and A.E. Charola, "The Physical Chemistry of Deteriorated Brick and Its Impregnation Technique," paper presented at the Congress for the Brick of Venice, October 22, 1979.

11. K. Lal Gauri, The Preservation of Stone, *Scientific American*, June 1978.

12. R.W. Drisko, An Introduction to Protective Coatings, *Public Works*, pp. 80–83 (August 1979).

Analytical Methods Related to Building and Monument Preservation

ISIDORE ADLER, SHELDON E. SOMMER,
RAPHAEL GERSHON, and JACOB I. TROMBKA

The most visible products of the weathering of stone materials are a consequence of the fragmentation and disintegration of mineral components. Somewhat less obvious are the dissolution of these minerals and subsequent formation of new compounds, frequently in the interstices, as a result of the action of chemical and biological agents. An early phenomenon that lends itself to study is the disruption of chemical bonds during physical and chemical disintegration and the formation of highly reactive surfaces. These reactions may include oxidation-reduction, disordering of the mineral structure, and ion-exchange processes, with the eventual formation of microlayers of poorly crystalline materials and microsystems of cracks and fractures with precipitated coatings, cements, and possible phase transformations as complicating factors.

The examination of these veneers presents problems that are well matched by the techniques utilized in bulk characterization, such as atomic absorption, X-ray fluorescence, and optical emission spectroscopy. With regard to surfaces and near surfaces (defined as 10 Å to a few micrometers in depth), one may consider a variety of techniques, some of which can be utilized in situ, offering the advantage of rapid and nondestructive analysis. The use of neutron–gamma techniques and reflection spectrophotometry are described as examples. Other techniques applied in the laboratory and that also require minimal sampling are electron spectroscopy, electron microprobe analysis, electron microscopy, and X-ray diffraction analysis. This paper examines the use of a number of these techniques, pointing out where a given method or combination of methods is most applicable and the way in which the results may be related to the weathering processes that are occurring.

Isidore Adler *is Professor, Departments of Chemistry and Geology, University of Maryland, College Park.* Sheldon E. Sommer *is Associate Professor of Geology, University of Maryland, College Park.* Raphael Gershon *is Student Assistant, Departments of Chemistry and Geology, University of Maryland, College Park.* Jacob I. Trombka *is Senior Scientist, Goddard Space Flight Center, Greenbelt, Maryland.*

INTRODUCTION

The disintegration and decomposition of stone materials—natural rock or mineral components and fabricated composites—often result in the formation of a veneer that differs from the original material in composition and texture. The new minerals produced by this weathering process are the result of physical, chemical, and biological reactions of carbonates, silicates, sulfides, or oxides with water and atmospheric gases. The typical products are often hydrated phases, such as clay minerals and iron and aluminum oxyhydroxides. In addition to this process, termed "hydrolysis," the oxidation of ferrous iron to ferric iron, and carbonation—chiefly the dissolution of limestone and marble by acidic waters—are major agents of rock weathering. These actions, coupled with ion exchange and physical and biological alteration, produce a marked change immediately below the stone–atmosphere interface.

Although the differential stability of components in stone materials depends on the complex interaction of various ambient materials with the primary and subsequently formed substances, general understanding of mineral degradation may be derived from a study of relative bond strengths. The removal of alkali and alkaline earth elements, resulting in the residual buildup of layers rich in silica, aluminum, and titanium, appears to be, for silicate rocks, a representation of the relative cation-oxygen bond strength. The various surface layers are rendered less stable by progressive bond rupture, and fragments of the mineral's structural framework are liberated in solution or otherwise altered. The weathering of many stone materials is so complex that there is little agreement on the mechanisms at work or on the methodologies best suited to such study.

The weathering of feldspar minerals, a major component of silicate-rich stone, has been a very active area of research in recent years. There are at least four different models of feldspar decomposition, i.e., for just one component of typical stone building material.[1] The models include: (a) the straightforward dissolution of the material, with the solubility controlled by the concentration of silica and alumina; (b) the production of a leached layer by the exchange of cations upward and downward through the interior of grains, in addition to solution at the interface; (c) the production of an amorphous precipitate rich in aluminum and silicon that is rate controlled and dependent on pH; and (d) the production of a crystalline phase dependent on solution composition and parent solid.

This brief summary of the possible analytical context strongly suggests that the proper methodology for the study of stone degradation is one that is capable of: (a) characterizing the surface or near-surface (tens of angstroms to hundreds of micrometers); (b) identifying amorphous and crystalline materials; (c) determining spatial changes in composition, i.e., chemical analyses for materials heterogeneous on a micrometer level; and (d) detailing the relative bond strengths as a function of physical and chemical alteration.

The analytical methods described below have been selected based on the above criteria. In addition, techniques are discussed that offer the advantage of in situ study for the characterization of alterations. These techniques may be used prior to, or perhaps in place of, destructive sampling of historic materials. We shall briefly list the principles and some examples of application.

X-RAY FLUORESCENCE SPECTROSCOPY

Various uses of X-ray fluorescence spectroscopy have been described.[2–4] Any process that produces inner-shell vacancies (i.e., ionization of an atom in its inner shell) will in turn produce characteristic X-rays. To create holes in an atom it is necessary to overcome in some fashion the binding energy of an electron in its particular shell. There are several ways of doing this. A target can be bombarded with electrons, energetic protons, alpha particles, or X-rays. As a case in point, if an X-ray photon has energy in excess of the binding energy of an electron in its shell, it will expel the electron from the atom by the photoelectric process, producing a vacancy and as a consequence an excited atom. The filling of this vacancy by outer-shell electrons as the atom returns to its ground state results in part in the emission of X-rays. Further, as the electrons from outer shells drop into the vacancies in the inner shells, new vacancies are produced and an electron cascade ensues. Electron transitions that end at the K shell produce a K spectrum. One can also expect to see L spectra, M spectra, etc. Examples of the possible transitions are shown in Figure 1. Note that transitions to outer shells produce a correspondingly increasing number of lines because of the greater number of possible transitions. Any particular transition results in a line whose energy, $h\nu$, is the difference between the binding energies of the two levels. These emitted lines are characteristic of the element.

Not every ionization results in the emission of a characteristic X-ray photon, however. There is in fact a very high probability of a radiationless transition in which the atom returns to its ground state by the emission of an electron known as the Auger electron. The

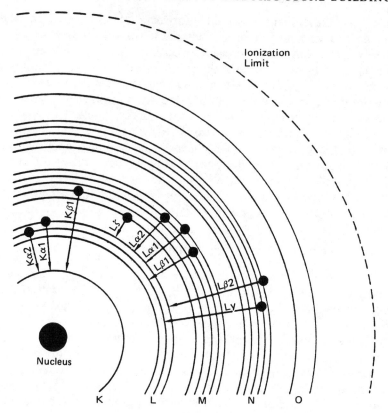

FIGURE 1   Possible electron transitions.

probability of this type of event increases markedly as the atomic number decreases. The Auger electrons also have characteristic energies, as we will see in the section Electron Spectroscopy–Chemical Analysis. Further, the photoelectron ejected initially also carries chemical information, since its maximum energy is equal to the difference between the energy of the exciting X-rays and the binding energy. Thus, in summary, bombardment by X-rays produces secondary X-rays, photoelectrons, and Auger electrons, all of which can yield information enabling us to identify an element and to determine its concentration and its chemical state.

The instrumentation used in the practice of X-ray fluorescence spectroscopy falls into two broad types, described as "wavelength-dispersive" or "energy-dispersive." In the wavelength-dispersive mode the various wavelengths produced in the sample are separated for measurement by diffraction from a large single crystal and then detected by a proportional, scintillation, or solid-state detector. In the energy-dispersive mode all the wavelengths are seen simultaneously by an energy-sensitive detector. The detector then produces pulses proportional in size to the incident energies. The pulses are then sorted on the basis of their heights by an electronic, window-type discriminator. Both modes have particular advantages. Wavelength-dispersive systems have the virtue of superior energy resolution, but the instrumentation is more complex mechanically, involving a precise crystal monochromator. The energy-dispersive systems are simpler and more efficient but are inferior in inherent energy resolution. The latter mode requires that the energy-separation problems be resolved by sophisticated software/computer methods.

Figure 2 shows both types of devices. Figure 2a is a plain view of wavelength-dispersive instrumentation. It consists of an exciting source (X-ray tube), collimators, an analyzing crystal, and a detector. The analyzer is based on Bragg's law, $n\lambda = 2d \sin \theta$, where n is the diffraction order, λ is the wavelength, d is the distance between the planes in the crystal, and θ is the angle of incidence or the diffraction angle. The expression shows that a given wavelength will diffract at a given angle depending on the d spacing of the crystal. In practice, the detector is made to rotate at twice the angular speed of the crystal. A given wavelength (corresponding to a particular element) will be detected as it satisfies Bragg's law. It has also been well established that, to a first order, the intensity of a line is proportional to concentration. Thus, we have the basis of an analytical method.

Figure 2b presents a line representation of an energy-dispersive system. As indicated above, such instrumentation is at least mechanically simpler than the wavelength-dispersive equipment, but it is electronically more complex. The output pulses of the detector are processed by a preamplifier and amplifier. These pulses are then sorted by a multichannel analyzer, which not only sorts the pulses by size but also delivers a number that is the sum of the pulses of a given size. Calibration involves relating pulse size to element. In the modern energy-dispersive analyzer, software programs in a dedicated computer identify the elements during the data-reduction phase. Of particular significance is the way in which this latter mode lends itself to in situ devices. There are in fact portable instruments commercially available for in situ analysis; they use radioactive sources to produce the X-rays.

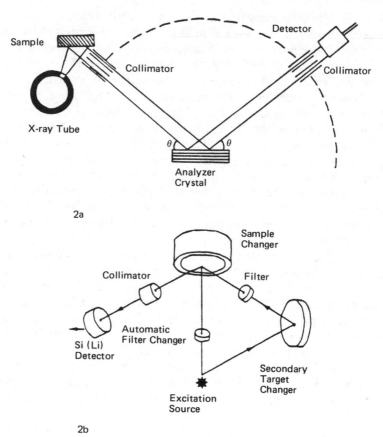

FIGURE 2 Representation of wavelength-dispersive and energy-dispersive equipment for X-ray fluorescence spectroscopy.

X-RAY DIFFRACTION

The power of X-ray fluorescence lies in its use for elemental analysis, whereas the analysis of crystalline phases falls within the province of X-ray diffraction. If one refers again to the Bragg expression for diffraction, $n\lambda = 2d \sin \theta$, the difference between the two techniques becomes clear. In the X-ray fluorescence mode the known values are the lattice dimensions of the crystal, d, and the Bragg angle, θ; λ, the unknown, is then simply determined and related to the element. In the diffraction case, a known X-ray wavelength is employed, and the diffraction angle, θ, is measured. These are then combined to determine the lattice parameters of the crystalline material that makes up the sample. Interpretations are drawn based on the values of the Bragg angles and the relative intensities of the various lines.

The instrumental arrangement is shown in Figure 3. The basic components consist of a source of X-radiation monochromatized by appropriate X-ray filters, the diffracting specimen, a radiation detector, a rate meter, and a recorder synchronized to the motion of the goniometer.

In a general way, any crystalline powder will produce a characteristic pattern. Such patterns are used for qualitative analysis, leading to the identification of the phase or compound. Specific identifications are usually made by reference to data in the Powder Diffraction File maintained by the American Society for Testing and Materials (ASTM). Given a mixture of crystalline materials, the resulting diffraction patterns will consist of superimposed patterns of the individual components. Interpretation is somewhat complicated, but X-ray diffraction is nevertheless useful for analyzing mixtures. It should be apparent that the use of techniques for a preliminary elemental analysis can be of great

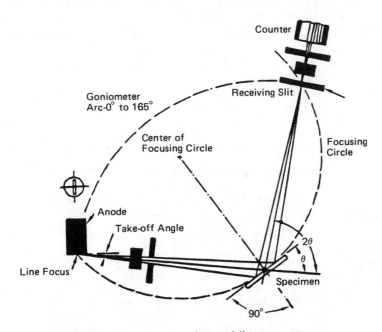

FIGURE 3 Representation of X-ray diffraction equipment.

value in supplying clues to the nature of the compounds. Finally, as in X-ray fluorescence, X-ray diffraction is nondestructive.

ELECTRON MICROPROBE AND SCANNING ELECTRON MICROSCOPE

The utilization of an electron beam focused on a small cross-sectional area of a sample allows for the spatial probing of composition and topography. The interaction of primary electrons with a sample produces signals—for example, X-rays, cathodoluminescence, back-scattered electrons, Auger electrons, and transmitted or absorbed electrons—that are related to elemental composition. There are also signals related to the topography of the surface, such as secondary electrons and, to a lesser degree, back-scattered electrons (Figure 4).

The version of an electron column instrument that has as its primary function the utilization of characteristic X-radiation produced by electron bombardment is termed an electron microprobe. This X-radiation yields compositional information from a spot as small as 1 μm in diameter and so may be used to determine variation in elemental content both in area distribution and within a surface layer whose depth approximates the diameter of the spot. A similar instrument is the scanning electron microscope (SEM).[5,6] Its primary function is to utilize the variation in secondary electron emission (electrons scattered

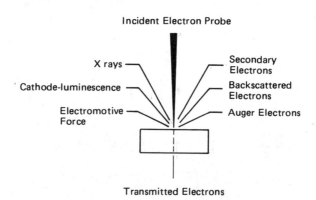

FIGURE 4   A beam of primary electrons, focused on a small cross-sectional area of a sample, produces a variety of signals related to the elemental composition of the sample.

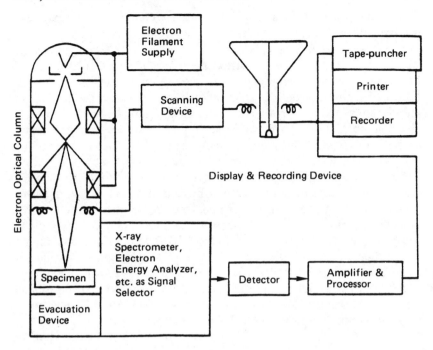

FIGURE 5 Typical layout of an electron microprobe analyzer.

by the surface with loss in energy) that occurs because of differences in surface topography as the electron beam sweeps in a raster (TV-type scan) across the sample surface.

The electron microprobe and the SEM were developed as separate instruments. Their similarities have been merged in modern instruments capable of performing both functions. A modern microprobe usually utilizes a crystal or wavelength spectrometer for X-ray identification, while an SEM utilizes an energy-dispersive (solid state) analyzer for X-ray identification. (See the section on X-ray fluorescence spectroscopy.) These are operational distinctions, because an instrument may be outfitted with either X-ray system. A microprobe is usually devoted to the highest quality X-ray analyses and is equipped to do optical microscopy concurrently with chemical analyses. An example of a typical microprobe layout is shown in Figure 5. A conventional SEM will differ: (a) in the type and number of electromagnetic lenses for focusing the electron beam, (b) in the absence of an optical microscope, and (c) in the use of an alternate X-ray detection system (if available). Samples must be ground and polished to a flat surface for quantitative analyses by the SEM.

The SEM is best utilized as a topographical analyzer of rough fracture surfaces, coupled with semiquantitative or qualitative elemental analyses. The electron microprobe is capable of determining all elements from boron through uranium, although analysis is usually limited to all elements above oxygen. The SEM most often analyzes elements above sodium, although the analyses typically are less accurate than with the microprobe. Either of these instruments is capable of resolving 50–100 Å in the secondary electron mode.

An important distinction should be noted between the electron spot size (or electron resolution) and the volume from which X-rays or other signals are being produced or detected (see Figure 6). Auger electrons typically are obtained from dimensions of tens of Å, secondary electrons from 50 to 250 Å, and X-rays from 1000 Å to a micron or more. This range in spatial resolution results in some signals (secondary electron and Auger) that have the same resolution as the primary probe and others (X-rays and back-scattered) that have poorer resolutions. Thus the location of elements in the sample as viewed by electrons does not coincide exactly with the source of the X-ray production.

The manner in which the X-rays are processed and converted to intensities is similar to the procedure detailed in the discussion of X-ray methods. The complex compositions of many building materials,

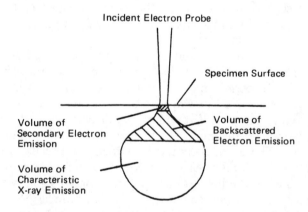

FIGURE 6   The size of the spot on which the electron beam is focused differs from the volumes from which the various signals are produced, with consequent differences in resolution.

especially silicates like granites or schists, require that a data-correction procedure be utilized to compensate for interelement and matrix effects arising from differential X-ray absorption and enhancement processes. These procedures are usually handled by an on-line computer to allow the operator to evaluate the data within minutes of the analyses.

The SEM is most suitable for studies where details of particle orientation and size as well as textural details, such as packing density, void space, recrystallization, or reprecipitation features, are beyond the reach of light microscopy. The SEM offers a 100-fold increase in depth of field over the light microscope, an increase in magnification of 50–100×, and a corresponding improvement in resolving power of 100×. The added attraction of performing energy-dispersive X-ray analyses on the rough sample is that qualitative elemental composition can then be used as an aid in detailing the characterization.

The electron microprobe is best suited to detailed quantitative analyses of flat surfaces, where no surface irregularities exist. The very powerful data reduction–correction procedures may then be utilized for a micrometer-level characterization. Both types of instruments allow for an X-ray map format, on which elemental distribution is displayed as a white–black dot matrix on a cathode-ray tube. Alternatively, a line scan may be used for the quantitative distribution of an element in a predetermined direction. This display, visual or printed, is well suited to chemical analyses along a transverse line from the outer portions of a stone sample to its interior.

Two lesser-known techniques in detailing the form and composition of stone materials are cathodoluminescence analyses and wavelength-shift effect. Cathodoluminescence (CL) refers to the light produced upon electron bombardment of certain materials, especially when activator ions such as Mn^{+2} are present. Many of these materials, including carbonates, silicates, and other building-stone materials, produce CL in sufficient quantity that details of fractures, recrystallization, and alteration invisible to optical or electron viewing are visible to the eye or to a suitable detector. The wavelength-shift effect refers to the slight shift in wavelength or energy of the X-ray emission of elements of low atomic number—e.g., silver, aluminum, sulfur, and phosphorus—as a function of their chemical or mineralogical environment. A sample of aluminum as Al_2O_3 has a measurable difference in shift of wavelength from that of aluminum in an aluminosilicate. This effect is of major use in the study of mortars and weathered surfaces where noncrystalline (to X-ray diffraction) or amorphous coatings defy phase characterization. The shifts can be related empirically to minerals whose shifts have been studied. This chemical–environment parameter yields information analogous to that obtained by ESCA (see following section).

ELECTRON SPECTROSCOPY–CHEMICAL ANALYSIS

Among the various instrumental techniques, one of the fastest growing is electron spectroscopy–chemical analysis (ESCA).[7] The method is built on the study of the energy distribution among the electrons ejected from a target material that is being irradiated by X-rays, ultraviolet radiation, or electrons. A convenient method for distinguishing the various kinds of electron spectroscopies is by the mode of excitation. The categories are X-ray photoelectron spectroscopy (XPS); ESCA ultra-violet photoelectron spectroscopy (UPS); or Auger spectroscopy, in which electron excitation is employed. Of the three types, ESCA has been perhaps the most used for chemical studies. The power of ESCA lies in its extraordinary sensitivity to surface chemistry. The method is sensitive to monolayers and involves distances of the order of angstroms. Further, the emerging electrons carry important information about such parameters as binding energies, charges, and valence states. A unique quality of ESCA is that it permits direct probing of the valence and core electrons. Figure 7 is a schematic illustration of the production of primary photoelectrons and Auger (secondary) electrons in an atom. The probability of photoelectron absorption depends on the energy of the incident photon and the atomic number of the element being irradiated.

To a first approximation the kinetic energy of the photoelectron is given by:

$$E_p = h\nu - E_b,$$

where E_p is the kinetic energy of the photoelectron, $h\nu$ is the energy of the incident photon, and E_b is the binding energy of the electron in its particular shell.

Thus, if the incident photons are "monoenergetic," the photoelectrons ejected from a given atomic shell will also be monoenergetic. For a given incident energy of the photons, the photoelectron spectrum will be characteristic, reflecting the various occupied electronic levels and bands in the material. It is necessary to emphasize, however, that the photoelectrons possess the characteristic energies as they leave the atom but that only a relatively small fraction of them emerge from a

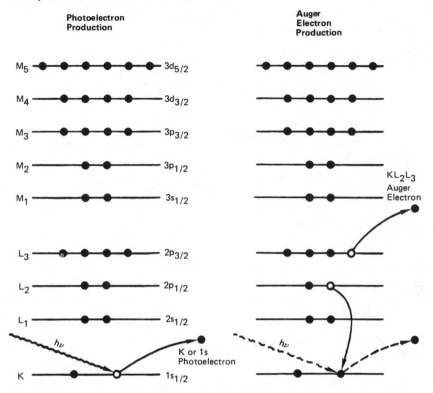

FIGURE 7 Schematic illustration of the production of primary photoelectrons and secondary (Auger) electrons in an atom.

target material with their energies undisturbed. This follows from the fact that electrons lose energy by a variety of processes as they leave a sample.

A typical arrangement for performing ESCA is shown in Figure 8. The necessary components include an X-ray excitation source (usually a X-ray tube containing a magnesium or aluminum target), the sample, an electron energy analyzer, and the appropriate electronics for pulse counting. Such instrumentation is available today in various commercial forms and in various degrees of sophistication. To summarize, ESCA is among the most powerful of the laboratory tools for the examination of surfaces.

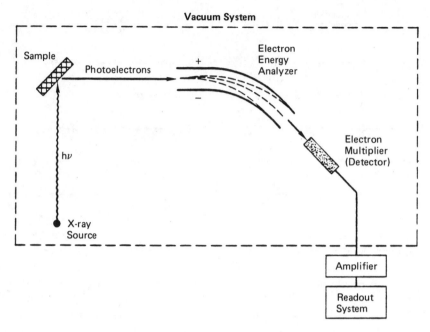

FIGURE 8   Typical instrumental arrangement for electron spectroscopy–chemical analysis.

IN SITU ANALYSES BY VISIBLE AND NEAR-INFRARED REFLECTANCE RADIOMETRY

The use of reflectance spectroscopy to study geological materials is well established.[7] This technique utilizes radiation reflected from a surface illuminated by the sun or an artificial source to record electronic (atomic) and vibrational (molecular) interactions at the surfaces of materials. The electronic processes associated with iron are of special interest because of this element's association with weathering and because the reflection features produced by the interaction of iron's d-shell electrons with its surroundings are within the detection capabilities of field instruments.

It is important to note that these in situ devices have been designed to operate only in the spectral regions accessible to sensors used by satellite and aircraft remote-sensing systems and so are limited to spectral bands not absorbed by the atmosphere.[8] However, these systems may be adapted to other spectral bands by utilizing other filters and detectors. In this manner it may be possible to measure spectral features assigned to the presence of silica or carbonate. The currently available systems allow us to distinguish some of the iron oxides and oxyhydroxides—e.g., goethite and hematite. Because the reflection spectra are often affected by particle size, we may be able to establish the relative size distribution of minerals as well. Of more immediate use is the radiometer's ability to detect the vibrational modes of the hydroxyl (OH) group, which is a major component of clay minerals and so may be used to identify various clays as well as some sulfate minerals.

Recent developments in the application of reflectance spectra to the weathering and alteration of rocks and minerals indicate that the ratios of various band intensities in the visible (electronic) and near-infrared (molecular) spectra allow for semiquantitative determination of various clay minerals, calcite (limestone/marble), iron minerals, and sulfates.[9] The spectra of clay minerals illustrated in Figure 9 have been divided into characteristic ratios—e.g., 2.0/2.20 μm and 2.20/2.35 μm— enabling mineral differentiation.

The hand-held radiometer is nearly the size of a small suitcase and is fully portable. The device may be set to a dual-beam mode for readout in a ratio format or to read in a multichannel mode of 3–20 different wavelengths. The capabilities of this type of instrument allow for nondestructive, field or laboratory measurement of the major minerals produced during the weathering/alteration of stone materials. Furthermore, the measurement is limited to near-surface penetrations— i.e., the weathered zone. An example of the change in the optical spectra caused by the thickness of the absorbing layer is illustrated in Figure 10. This change in reflectance/absorption may be utilized to determine the depth of alteration, leaching, or weathering. As a function of mineralogy (i.e., wavelength), the spectra are representative of the upper 20–50 μm of the sample.[9]

NEUTRON–GAMMA TECHNIQUES

The use of prompt neutron–gamma techniques is well established in geochemical exploration and environmental monitoring. Prompt neutron techniques have been proposed and used for such applications as borehole logging[10] and the detection of pollutants in river and sea beds.[11] The prompt neutron–gamma method involves the measurement of gamma rays that result from the interactions of neutrons with the material under analysis. Fast neutrons (energy >1 MeV) can interact by scattering inelastically from a nucleus, thereby producing a nucleus in an excited state. Subsequent deexcitation results in the

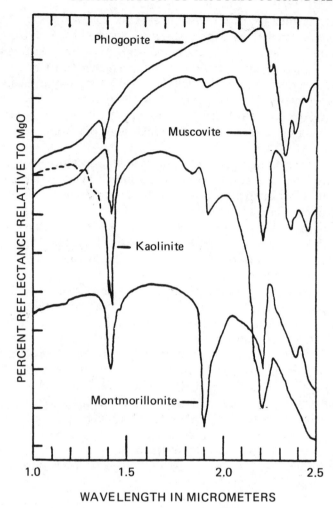

FIGURE 9 Spectra of clay minerals determined by reflectance radiometry. Spectra are superimposed in this figure with indicated spacings of 10 percent reflectance to MgO. Source: G. R. Hunt in *Remote Sensing in Geology*, B. S. Siegal and A. R. Gillespie, eds. John Wiley: New York, 1980.

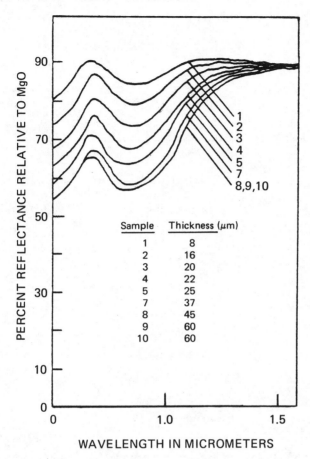

Sample	Thickness (μm)
1	8
2	16
3	20
4	22
5	25
7	37
8	45
9	60
10	60

PERCENT REFLECTANCE RELATIVE TO MgO

WAVELENGTH IN MICROMETERS

FIGURE 10 Reflectance spectra of 25 percent goethite and 75 percent kaolinite as a function of the thickness of the absorbing layer.

emission of characteristic gamma rays. Thermal neutrons can be captured by a nucleus, which increases its atomic mass by one and in the process is left in an excited state. The return to the ground state is accompanied by the emission of a characteristic gamma ray. The gamma-ray flux measured by a detector depends on the spatial and energy distribution of the neutrons and the location of the detector relative to the source. Figure 11 summarizes both modes of analysis. A possible instrumental configuration is shown in Figure 12.

An example of the great amount of data available in a spectrum

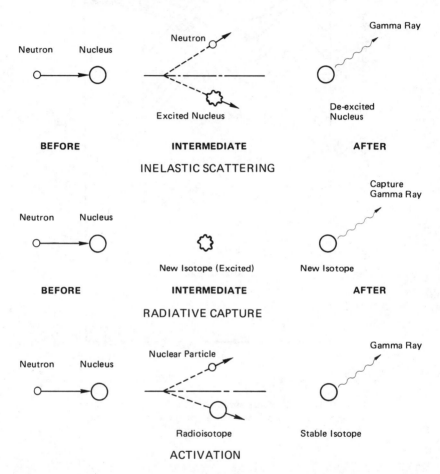

FIGURE 11 Neutron–gamma methods of analysis involve detection of characteristic gamma rays emitted as a result of inelastic scattering of fast neutrons or radiative capture of thermal neutrons.

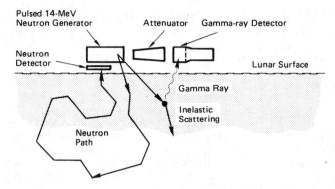

FIGURE 12 Possible instrumental configuration for analysis by the neutron–gamma method.

accumulated in about an hour is shown in Figure 13. This is a spectrum taken of wet soil in an anticoincidence mode. It was possible to identify hydrogen, oxygen, silicon, iron, aluminum, titanium, sodium, calcium, and potassium. Most of the lines were due to neutron capture. It was also possible to identify lines resulting from neutron activation. While this method has not been applied specifically to the problems being discussed at this conference, it appears to hold promise. The major constraints to be considered are that the methods provide strictly el-

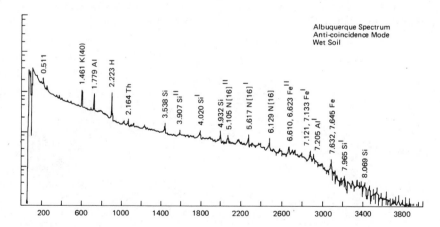

FIGURE 13 The extensive spectral data shown in the figure were obtained in about an hour using the neutron–gamma method on a sample of wet soil.

emental data, and the analysis represents a bulk sample. One important positive factor is that the methods are capable of yielding information about the hydrogen content.

REFERENCES AND NOTES

1. See, for example, the following recent studies: Petrovic et al. (1976) Rate control in dissolution of alkali feldspars I. Study of residual feldspar grains by X-ray photoelectron spectroscopy, 40(5), 537; Busenberg (1978) The products of the interaction of feldspars with aqueous solutions at 25°C, 42(111), 1679; Holdren and Berner (1979) Mechanism of feldspar weathering I. Experimental studies, 43 (8), 1161–1187; Tsuzuki and Suzuki (1980) Experimental study of the alteration of labradorite in acid hydrothermal solutions, 44(5), 673. All articles are from *Geochemica Acta*.

2. Adler, I. (1966), X-Ray Emission Spectrography in Geology, Elsevier, N.Y.

3. Liebhafsky, H.A., Pfeiffer, H.D., Winslow, F.H. and Zemany, D.D. (1960) X-ray Absorption and Emission in Analytical Chemistry, John Wiley, N.Y.

4. Auger, D. (1925) Secondary β-rays produced in O gas by X-rays. *Compt. Rend.* 180:65–68.

5. Smith, D.G.W. (1976) Short Course in Microbeam Techniques, Mineralogical Association of Canada, Toronto, Ontario.

6. Goldstein, J.I., et al. (1975) Practical Scanning Electron Microscopy, Plenum Press, N.Y.

7. Yin Lo I, and Adler, I. (1978) Electron Spectroscopy, Instrumental Analysis, 418–442, Allyn and Bacon, N.Y.

8. Tucker, C.J., et al. (1980) NASA Tech. Mem. 80641, NASA–GSFC. Barringer Company Sales Literature, Denver, Colorado.

9. Hunt, G.R. (1961) Spectral studies of particulate minerals in the visible and near infrared, *Geophysics*, 42(3), 501.

10. Hertzog, R.C., Plasek, R.E. (1979) Neutron excited gamma-ray spectrometry for well logging, *IEE Trans. Nucl. Sci.*, NS-26, p. 1558.

11. Johnson, R.G., Evans, L.G., Trombka, J.I. (1979) Neutron–gamma techniques for planetary exploration, *IEE Trans. Nucl. Sci.*, NS-26, p. 1574.

Wet and Dry Surface Deposition
of Air Pollutants and Their Modeling

BRUCE B. HICKS

The net rate of delivery of trace gases to receptor surfaces is largely determined by the chemical affinity of surface materials for the gas in question. If molecules of the gas are captured efficiently or react quickly upon contact with the surface, then high surface flux densities can be expected. Large particles are deposited by gravitational settling and by inertial impaction; the efficiency of their capture depends on their shape and the structure of the surface at the point of impact. Small, submicron particles have difficulty penetrating the quasilaminar air layer adjacent to smooth surfaces, but once they contact the surface they are efficiently retained by van der Waals forces. All particles are susceptible to electrostatic forces that will encourage deposition if either the particles or the receptor surfaces carry an electrical charge. The presence of temperature and humidity gradients near the surface can also promote or hinder the deposition of particles. Most of these matters have been investigated in studies of deposition to relatively uniform surfaces of pipes or plates in wind tunnels. Extrapolation to the real-world case of complicated surface shapes is sufficiently uncertain that quantitative statements cannot be made. The role of rainfall and other kinds of atmospheric precipitation is equally complicated. Current ecological concern about the acidity of rain has focused attention on adverse effects associated with precipitation chemistry, but it must be recognized that rainfall provides a natural cleansing mechanism in many instances. In highly polluted areas, it is possible that the major effect of rainfall will be to remove some previously deposited pollutants from exposed surfaces and promote the subsequent deposition of soluble gases and small particles to those areas (such as crevices) that remain moist.

Bruce B. Hicks *is Director, NOAA Atmospheric Turbulence and Diffusion Laboratory, Oak Ridge, Tenn.*

This work was supported in part by the Multistate Atmospheric Power Production Pollution Study and sponsored by the U.S. Environmental Protection Agency.

183

There has been considerable recent work on the transfer of air pollutants to receptor surfaces. Much of this work has been associated with concern about potential ecological effects of the increases in atmospheric sulfur loading expected to accompany an increase in the use of coal as an energy source. Current fears about acid rain have concentrated attention on chemical deposition by precipitation, but there is a continuing awareness that dry deposition processes are capable of delivering similar quantities of material even to areas fairly distant from pollution sources. With near sources, such as within cities where pollution levels are high, we must expect dry mechanisms to deliver at least as much material to exposed surfaces as wet, especially when the surfaces in question are sloping or are somehow protected from the direct impact of precipitation.

The results of modern research on ecological factors associated with chemical deposition are not usually transferable to the case of stone weathering because the ecological work places strong emphasis on matters related to biology. However, a small component of these studies seeks to identify and formulate the mechanisms that control the rates of deposition of airborne pollutants. This work combines theoretical and laboratory research with field investigations of pollutant fluxes to provide a comprehensive understanding of the processes that determine the dry fluxes of many trace gases and small particles to uniform, natural surfaces. In the present context of deposition to stonework, the recent ecologically oriented work allows us to reconsider some of the formulations developed in earlier chemical engineering studies of the deposition to flat plates and to the surfaces of pipes.

Likewise, recent work on the chemistry of rainfall has tended to concentrate on its acidic properties and their possible changes with both time and space, since these factors are of definite ecological importance. These studies have provided greatly improved understanding of the processes that combine to produce polluted rain and have given workers a much better feel for the natural variability of precipitation chemistry. But before discussing details of the wet and dry deposition processes that are capable of delivering pollutants to exposed stonework, it is useful to consider the mechanisms that contribute to deterioration and hence to identify the specific deposition phenomena that are likely to be most important. The mechanisms are:

- *Physical Mechanisms*   The presence of water at the surface is known to be a key factor in promoting the fracturing and erosion of stone. Water penetrates pores and cracks and causes mechanical stresses both by freezing and by the hydration and subsequent crystallization of salts.[1,2,3,4]

- *Chemical Mechanisms* Some deposited chemical agents will react with stone surfaces. Sulfur compounds have been indicted as the most critical factors in this regard,[5] mainly because they are often acidic and can have high concentrations in city and suburban air; but nitrogen compounds should be considered as well. Fluxes of trace gases (e.g., sulfur dioxide) can be high, especially when promoted by biological activity like that mentioned below. Dissolution by chemical reaction with contaminants contained in precipitation is one of the most familiar eroding processes, particularly in the case of carbonaceous stone. Details of the chemical reactions involved are well documented.[6]
- *Biological Mechanisms* Many different biological factors have been shown to be important. Growths of lichens, mosses, algae, mold, fungi, and bacteria are capable of promoting at least surface deterioration.[7] Some bacteria can synthesize sulfuric (or nitric) acid from airborne sulfur dioxide (or nitrogen oxides). Guano contains phosphoric acid, which can also cause considerable damage.[3]

In light of the above comments, it appears desirable to focus present attention on the deposition of sulfur dioxide and small (potentially acidic) particles, on the condensation of water at the surface and at already deposited particles, and on the characteristics of pollutants delivered in precipitation.

DRY DEPOSITION

A pollutant in air near a surface will be transported to the immediate vicinity of the surface by average winds and turbulence. This process is usually rapid; only at night, when conditions become very calm, can pollutant uptake rates be limited by low turbulence. As a pollutant approaches the surface, molecular (or Brownian, in the case of particles) diffusion becomes increasingly important. Brownian diffusivity can be so low, however, that aerosol particles have difficulty penetrating the quasilaminar layer adjacent to the surface. Once a pollutant particle or molecule contacts the surface, it is not necessarily captured (although van der Waals forces are usually considered sufficient to capture particles).[8] Thus, there is a surface resistance that quantifies the absorption of trace gases or the retention of particles at the surface. Once material is deposited, chemical reactions can impose further variability on the overall uptake characteristics and are likely to be especially important if the surface is wet. The entire process of particle deposition will be modified if particles carry an electrostatic charge.

Much of what we know about nongravitational deposition of gases and small particles to surfaces follows from studies of transfer to the walls of pipes from fluids flowing through them. These studies have shown, for example, that the transfer coefficient associated with diffusion through the quasilaminar layer in contact with smooth surfaces can be conveniently formulated in terms of the diffusivity of the quantity in question (in most literature, nondimensionalized as the Schmidt number, $Sc = v/D$, where v is viscosity and D is the pollutant diffusivity) and the friction velocity, u_*. The conductance, or transfer velocity across the quasilaminar layer is proportional to u_*; the constant of proportionality is usually written as a quantity B that is then directly dependent on Sc. Figure 1 shows the results of several experiments which indicate that $B \sim Sc^{-2/3}$.

The similarity between deposition to flat smooth surfaces in con-

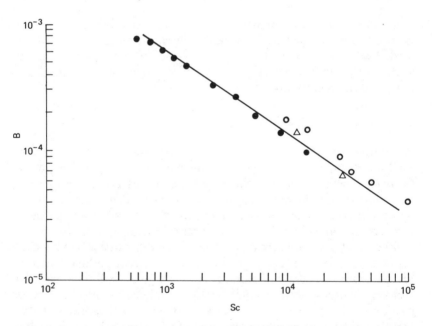

FIGURE 1 Variation of the surface boundary layer property B (see equation 1) with Schmidt number for transfer to smooth flat surfaces (after Lewellen and Sheng, 1980). Data are derived from Harriot and Hamilton (1965; open circles), Hubbard and Lightfoot (1966; triangles), and Mizushina et al. (1971; solid circles).

trolled circumstances and to stone surfaces in natural conditions may be somewhat limited. However, in wind tunnel studies they appear to be in general agreement. Moller and Schumann, for example, find close to a $Sc^{-2/3}$ dependence in the case of small-particle transfer to water surfaces in wind tunnels.[9] Figure 2 presents a familiar set of wind tunnel observations of the deposition velocity ($v_d = F/C$, where F is the flux density and C is the airborne concentration of the pollutant) to horizontal, flat surfaces, as a function of particle size. For particles sufficiently small that gravitational settling is insignificant, these results are dominated by near-surface, quasilaminar behavior such as that seen in Figure 1. Indeed, the line drawn on the left in Figure 2 has a slope compatible with that in Figure 1. The corresponding expression is:

$$v_d = u_\star B = A\ u_\star Sc^{-2/3}, \tag{1}$$

where A is a constant. There are, however, inconsistencies between the sets of results that cause some loss of confidence in generalizing expressions of this kind. First, it appears that the precise value of the exponent is not known. While studies of trace gas transfer[10] and particles[11] agree in the relevance of a $-2/3$ power law relationship, a survey by Brutsaert indicates exponents ranging between -0.4 and -0.8.[12]

Second, the value of the numerical constant A in equation 1 appears quite uncertain. The line drawn through the data of Figure 1 corresponds to $A \simeq 0.06$, yet agreement with the small-particle data of Figure 2 seems to require $A \simeq 0.6$. These values span the result recommended by Wesely and Hicks for the case of sulfur dioxide fluxes to fibrous, vegetated surfaces.[13] They suggest relations equivalent to $A \simeq 0.2$, as was indicated by earlier experiments conducted by Shepherd.[14]

Regardless of the uncertainties about the detailed formulation of deposition through the quasilaminar layer, it is clear that large particles will penetrate it less easily than smaller ones, unless influenced by gravity or some other process (such as inertial impaction). Figure 2 refers to the special case of smooth horizontal surfaces. At the right hand side of the diagram, observations conform with the predictions of Stokes-law settling, although with some enhancement because of inertial impaction.[8] Generalization of these results to sloping surfaces is not a trivial exercise, although calculations based on horizontally projected areas might provide acceptable estimates of gravitational settling rates in some situations. The contribution by inertial impac-

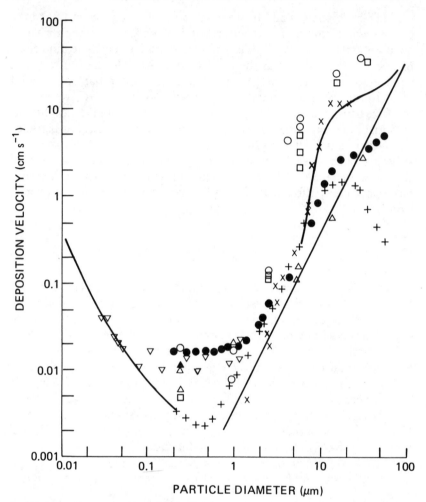

FIGURE 2　Particle-size dependence of the deposition velocity to relatively smooth horizontal surfaces. The line at the left represents a Schmidt-number relationship like that of Figure 1. At the right is the expected Stokes-law relationship, with a curve drawn by eye to draw attention to the enhancement due to inertial impaction. Crosses indicate results concerning aerodynamically smooth surfaces: vertical crosses to filter paper (Clough, 1973) and diagonal crosses to glass (Liu and Agarwal, 1974). Dots apply to artificially roughened surfaces (Clough, 1973). The remaining data refer to water surfaces, for friction velocities of 40 cm/s (Moller and Schumann, 1970; inverted triangles) and 11, 44, and 117 cm/s (Sehmel and Sutter, 1974; triangles, squares and circles).

tion also cannot be easily calculated, since it will be influenced greatly by local flow distortion and microscale roughness characteristics.

Recent work has shown that most aerosol acidity is associated with small particles of the "accumulation" size range, mostly between 0.2 and 1.0 μm diameter. In the context of the dry deposition of acidic particles to smooth surfaces, therefore, gravitational settling and inertial impaction become less important concerns than diffusive transport and surface retention. Diffusion has been considered above; near-surface transport is conveniently formulated in terms of the Schmidt number. If particles are efficiently captured on contact with the surfaces, then available information indicates that equation 1 provides a way to evaluate how deposition will vary with particle size.

The above arguments concerning laminar layers and their effect on the transfer of particulate pollutants are relevant when the surface involved is homogeneous and smooth. However, when it is roughened, the barrier presented by the quasilaminar layer is likely to be penetrated, and transfer rates might be considerably enhanced. In particular, large values of v_d might be expected when surface discontinuities and sharp irregularities occur with characteristic scales greater than the scale thickness of the diffusive layer. The depth of this layer is usually assumed to be determined by viscosity and the friction velocity; classical studies of flow over sand in wind tunnels indicate layer thicknesses of the order of 50 μm in moderate velocities (wind speed of a few m s^{-1}). Surface roughness elements of this characteristic size should therefore be suspected as sites for preferred deposition, especially if they are sharp and irregular.

PHORETIC EFFECTS

Transfer of both gases and particles can be influenced by evaporation from a wetted surface or by condensation. The role of condensation is of special interest, since as moisture is deposited at the surface, there is a mean flow of air to replace the condensed vapor, and net deposition velocities of all pollutants are increased accordingly. The magnitude of this Stefan flow, v_s, is readily calculable as:

$$v_s = (m_a/m_w) \cdot (E/\rho), \tag{2}$$

where E is the evaporation rate (in kg m^{-2}s^{-1}), ρ is air density, m_a is the molecular weight of air, and m_w is that of water. At standard temperature and pressure, the overall deposition velocity is increased by about 0.005 cm s^{-1} for every 100 W m^{-2} of latent heat transferred by condensation. The maximum rate of dewfall to vegetated surfaces is known to be about 45 W m^{-2} (equivalent to about 0.07 mm h^{-1})[21], so that the increase in v_d is unlikely to exceed 0.003 cm s^{-1}. Even lower values should apply in the case of statues and monuments, but these might still be of similar magnitude to the values predicted for the transfer of small particles to smooth surfaces, as considered in Figure 2.

In daytime, evaporation from wet surfaces will tend to protect them from pollutant deposition. When liquid water is present at the surface, evaporation rates are controlled by the availability of heat, primarily from insolation, and can easily be more than an order of magnitude greater than the condensation rates considered above. When evaporation is proceeding strongly, Stefan flow might provide a barrier against the deposition of those pollutants (especially small particles) having deposition velocities normally less than about 0.03 cm s^{-1}. Such strong evaporation rates are not infrequent, but wet stone surfaces will be rapidly dried and the effect cannot persist for long periods.

Stefan flow affects particles and gases alike; however, there are related mechanisms that act primarily on particles. These phoretic forces result from the response of particles to the impaction of air and water molecules in the presence of temperature and humidity gradients. The effect is to drive particles toward cold or evaporating surfaces.

Friedlander shows[8] that the thermophoretic velocity increment can be expressed as:

$$v_t \simeq -0.3 \ (v/T) \cdot \nabla T, \qquad (3)$$

where T is absolute temperature. The negative sign indicates the counterflux direction of the imposed motion (away from warm surfaces), and the constant 0.3 is actually a slight function of particle size (0.33 for 0.1 μm, 0.28 for 0.3 μm, and 0.16 for 1.0 μm). It is informative to introduce the sensible-heat flux H and to rewrite equation (3) as

$$v_t \simeq -0.3 \ Pr \ H/(\rho c_p T) \qquad (4)$$

where Pr is the Prandtl number, v/D_T, analogous to the Schmidt number in equation 1; D_T is thermal diffusivity, ρ is air density, and c_p is the specific heat of air at constant pressure. Equation 4 indicates an additional velocity increment amounting to about 0.0065 cm s^{-1} for every 100 W m^{-2} of sensible-heat flux. At night, H rarely exceeds this value and is typically -10 to -20 W m^{-2}, so that even though deposition is enhanced, it is only by a small amount. In daytime, however, heat fluxes can exceed 500 W m^{-2}, imposing a considerable barrier against particle deposition.

A similar phoretic force is exerted on particles by water molecules diffusing past them. Whereas the Stefan velocity considered above was a consequence of a mean displacement of the gas by evaporating water molecules, diffusiophoresis is a result of a net flux of water molecules past the particle in question and of a flux of heavier air molecules to replace them. The resulting velocity is sometimes combined with and often confused with the Stefan flow described earlier. The detailed investigation presented by Goldsmith and May shows diffusiophoresis to be far less important than either Stefan flow or thermophoresis in the circumstances of interest here.[22]

A word of caution seems appropriate at this stage. Many of the equations given above assume the existence of a laminar layer in contact with the surface, a situation that appears highly unlikely unless the surface under consideration is remarkably smooth and uniform (which may well be so when a stone surface is new and polished). Thus, we should avoid the use of these relations to quantify deposition with precision, but instead should employ them to identify critical properties and to determine the circumstances in which deposition will be greatest.

WET DEPOSITION

Recent interest in the quality of precipitation (especially its acidity) has resulted in a wealth of data on the chemical composition of rain and snow. Rainfall itself is a highly variable quantity that is known to conform quite closely to a log-normal frequency distribution. Its chemical characteristics are even more varied, but analyses of data obtained by the Multistate Atmospheric Power Production Pollution Study (MAP3S) show that pollutant concentrations in precipitation events also tend to follow the expected log-normal distribution.[23] Table 1 summarizes some of the results of the MAP3S survey, and combines them with data derived from sampling conducted by the Department of Energy Environmental Measurements Laboratory (DOE/EML).[24] The intent is to extend the MAP3S conclusions to suburban and city areas; the MAP3S sites are carefully selected to be well away from such areas and are intended to be indicative of regional-scale characteristics rather than the local variations that are likely to be mainly of interest here.

Hydrogen, sulfate, and nitrate ion concentrations are selected for consideration in Table 1. The rural sites in the northeastern United

TABLE 1 Long-Term Mean Concentrations of Hydrogen, Sulfate, and Nitrate Ions in Precipitation Collected at Selected Sites

	pH	H^+ $\overline{C}$	s_c	$SO_4^=$ $\overline{C}$	s_c	NO_3^- $\overline{C}$	s_c
Rural sites							
Chester, N.J.[a]	4.1	75		73		36	
Ithaca, N.Y.[b]	4.1	81	1.9	60	2.2	32	2.0
State College, Pa.[b]	4.1	79	2.2	62	2.4	39	2.2
Charlottesville, Va.[b]	4.1	72	2.1	61	2.2	28	2.3
Miami, Ohio[b]	4.2	65	1.9	65	1.8	29	1.9
Urbana, Ill.[b]	4.4	43	3.5	71	1.8	30	1.8
Beaverton, Oreg.[a]	5.5	3		20		7	
Urban and city sites							
Argonne, Ill.[a]	4.6	25		105		32	
New York, N.Y.[a]	3.9	130		150		40	

NOTE: Concentrations are in microequivalents per liter. $\overline{C}$ is the mean concentration, and s_c is the appropriate standard deviation. A log-normal distribution is assumed, so that s_c applies to the normal distribution of a variable $x = ln\ C$.
[a] Data from Feely and Larsen (1979).
[b] Data from the MAP3S network (MAP3S, 1981).

States appear to yield similar results: Hydrogen and sulfate ion concentrations are in the range 60 to 80 microequivalents per liter, and nitrate is 30 to 40. The Chester site of the DOE/EML network is well away from upwind sources and does indeed provide results that agree with the MAP3S data in the Northeast; hence the two data sets appear to be compatible, even though DOE/EML provides monthly averages only, whereas MAP3S concentrates on events. The Urbana data show evidence that midwestern rural precipitation chemistry is somewhat different from that farther east; while sulfate and nitrate concentrations appear much the same, hydrogen ion concentrations seem to be substantially lower. The Beaverton, Oregon, data extend this trend to the West Coast, where the precipitation is cleaner in all aspects.

The New York City data were obtained at a central rooftop location. Hydrogen ion and sulfate concentrations are about double the values typical of the Northeast as a whole, but nitrate concentrations are relatively unaffected. The Argonne, Illinois, data were obtained at a suburban site about 40 km upwind of downtown Chicago, but in an area influenced by considerable industrial activity. The Argonne nitrate values seem unaffected, but the sulfate concentrations are intermediate between regionally characteristic values and the New York City maximum. Hydrogen ion concentrations at Argonne seem strangely low, but are suspected of being influenced by soil-derived material capable of neutralizing the more acidic constituents. The Urbana data are thought to be similarly affected, but clearly to a lesser extent.

The MAP3S data provide excellent evidence that the log-normal frequency distribution usually associated with the precipitation process itself also provides a good representation of the chemical concentration data. The standard deviations listed in Table 1 show remarkable uniformity. With the sole exception of the Urbana hydrogen ion data, values range between 1.8 and 2.4, and the average is 2.0. Thus the shape of the frequency distribution is fairly constant and can therefore be extended with some confidence to the case of cities. In this way it is possible to estimate the probability of encountering exceedingly acidic rainfall when only its long-term average characteristics are known, and it is similarly possible to estimate the frequency distributions of concentrations of different chemical species. For central New York City, for example, Table 1 indicates that the probability of any single rainfall event producing a pH less than 3.0 is about 15 percent.

It might be noted that the standard deviations in Table 1 indicate a fairly constant spread of pH values at every site except Urbana. In general the standard deviation of the pH of event precipitation is about 0.9 (1.5 at Urbana). Since pH depends on the logarithm of the hydrogen ion concentration, the frequency distribution of event pH values will be close to Gaussian.

It is evident in Table 1 that even polluted rain has very low ionic strengths. A complication that is not evident in the table is that the greatest ionic strengths tend to be associated with the smallest rainfalls, so that high concentrations need not necessarily suggest large doses of chemical contaminants. Moreover, the probability of an extremely acidic event appears to be quite low, even in city and suburban environments. These considerations combine to suggest that deterioration by the action of chemicals in precipitation might not be as important as that caused by the hydration and mobilization of surface materials already deposited. In some situations the net washing effect might even be beneficial. It is certainly clear that local variations of precipitation chemistry can be large and that generalizing is bound to be dangerous.

AIR CONCENTRATIONS

The mechanisms that deposit pollutants constitute only the final step in a chain of events that transport and transform them between sources and receptors. The mechanisms that combine to result in the dry deposition of gaseous pollutants are likely to be most effective when the air is turbulent (i.e., mainly in daytime) and when the surface is moist and therefore acting as a good sink for soluble gases. For particles, dry deposition is mainly limited by transport very near the surface and may well be greatest when water is condensing. But none of these mechanisms will result in a large flux of pollutants unless sufficient concentrations of the pollutant are accessible in air near the surface.

Many variables contribute to the diurnal cycle in concentration of any air pollutant. The highest concentrations can occur at night in some cases and during the daytime in others. Variability of this kind is greatly influenced by the spatial distribution and height of emission sources. It is also greatly influenced by whether relatively undiluted pollutant plumes can be carried near the surface by atmospheric mixing (fumigation) or by some local interference with flow patterns, such as might be caused by a large building, a hill, or rows of trees. Thus there is no general consensus that sulfur dioxide, for example, will peak in the daytime, although an early morning peak associated with fumigation is frequently observed.

Superimposed on any regular cycle of this kind will be a random variability whose magnitude and frequency distribution will vary greatly with time and location. The temporal variability in air pollution concentrations is a well-known feature that emphasizes the need to obtain relevant data by experiment whenever specific areas of interest can be identified. The need is amplified by the additional uncertainty associated with spatial differences, especially within cities or in areas affected by local traffic. Furthermore, it is evident that more detail is required than is obtained in most pollution monitoring programs, since the deposition processes vary with the time of day. It would be difficult to interpret daily averaged concentration data, whereas averaged daily cycles and frequency distributions would be quite informative.

CONCLUSIONS

Both wet and dry deposition of pollutants can cause significant deterioration of exposed stonework. Wet deposition imposes sudden but infrequent doses of pollutants, most of which will be in dilute solution. Concentrations will vary widely both in time and in space, but as a rule of thumb the pH will be roughly normally distributed, with a standard deviation of about 1.0. It is obviously possible to protect exposed surfaces from the direct effect of precipitation, but it is not immediately clear that the use of shelters will generally be beneficial.

Dry deposition is a slower but more continuous process than wet deposition, and it is always possible that incident precipitation will wash off material previously deposited by dry processes. However, in cold weather the mechanical effects associated with repeated freezing and thawing of water are likely to overwhelm all other factors.

Both dry and wet fluxes will be greatest when air concentrations of pollutants are high. Although the relationship between air concentrations and the chemical composition of precipitation is exceedingly complicated, rates of deposition by dry mechanisms are intimately related to air quality in the immediate vicinity of receptor surfaces. Regarding dry deposition to exposed stonework, the following additional points seem clear:

• In daytime, particle fluxes will be greatest to the coolest parts of exposed surfaces.

• Both particle and gas fluxes will be increased when condensation is taking place at the surface and decreased when evaporation occurs.

• If the surface is wet, impinging particles will have a better chance of adhering, and soluble trace gases will be more readily captured.

• The chemical nature of the surface is important; if rates of reaction with deposited pollutants are rapid, then surfaces can act as nearly perfect sinks.

• Biological factors can influence uptake rates by modifying the ability of the surface to capture and bind pollutants.

• The texture of the surface is important. Rough surfaces will provide better deposition substrates than smoother surfaces and will permit easier transport of pollutants across the near-surface quasilaminar layer.

Finally, it should be emphasized that the present state of knowledge regarding pollutant uptake by surfaces of any kind is rather rudimentary. Nevertheless, important processes can be identified with some confidence. While the rates of uptake cannot be predicted at all closely, the circumstances under which the greatest fluxes occur can be determined. In particular, some surface properties that are likely to cause locally enhanced deposition can be identified, and hence areas that are potentially at risk can be singled out.

REFERENCES

1. Winkler, E.M., and E.J. Wilhelm. 1970. Saltburst by hydration pressures in architectural stone in urban atmosphere, *Geol. Soc. Am Bull.* 81:576–572.

2. Winkler, E.M., and P.C. Singer. 1972. Crystallization pressure of salts in stone and concrete, *Geol. Soc. Am. Bull.* 83: 3509–3514.

3. Fassina, V. 1978. A survey of air pollution and deterioration of stonework in Venice, *Atmos. Environ.* 12:2205–2211.

4. Gauri, K.L. 1978. The preservation of stone, *Sci. Am.* 238: 196–202.

5. Yocom, J.E., and R.O. McCaldin. 1968. Effects of air pollution on materials and the economy. In *Air Pollution* (A.C. Stern, ed.) Academic Press: New York.

6. Keller, W.D. 1977. Progress and problems in rock weathering related to stone decay In *Engineering Geology Case Histories Number 11*. Geological Society of America.

7. Torraca, G. 1977. Brick, adobe, stone, and architectural ceramics: Deterioration process and conservation practices, reprinted in *Papers in Atmospheric Science* (A. Mestitz and O. Vittori, eds.) Consiglio Nazionale delle Richereche: Italy.

8. Friedlander, S.K. 1977. *Smoke, Dust and Haze Fundamentals of Aerosol Behavior.* John Wiley: New York.

9. Moller, U. and G. Schumann. 1970: Mechanisms of transport from the atmosphere to the Earth's surface, *J. Geophys. Res.* 75: 3013–3019.

10. Deacon, E.L. 1977. Gas transfer to and across an air–water interface, *Tellus* 29: 363-374.

11. Lewellen, W.S., and Y.P. Sheng. 1980. *Modeling of Dry Deposition of SO_2 and Sulfate Aerosols.* Electric Power Research Institute Report EA-1452.

12. Brutsaert, W. 1975. The Roughness length for water vapor, sensible heat, and other scalars, *J. Atmos. Sci.* 32: 2028–2031.

13. Wesely, M.L., and B.B. Hicks. 1977. Some factors that affect the deposition rates of sulfur dioxide and similar gases on vegetation, *J. Air Pollut. Control Assoc.* 27: 1110–1116.

14. Shepherd, J.G. 1974. Measurements of the direct deposition of sulphur dioxide onto grass and water by the profile method, *Atmos. Environ.* 8:69–74.

15. Harriot, P., and R.M. Hamilton. 1965. Solid–liquid mass transfer in turbulent pipe flow, *Chem. Eng. Sci.* 20:1073.

16. Hubbard, D.W. and E.N. Lightfoot. 1966. Correlation of heat and mass transfer data for high Schmidt and Reynolds numbers, *I/EC Fundamentals* 5:370.

17. Mizushina, T., F. Ogino, Y. Oka, and H. Fukuda. 1971. Turbulent heat and mass transfer between wall and fluid streams of large Prandtl and Schmidt numbers, *Inter. J. Heat and Mass Transfer* 14:1705–1716.

18. Clough, W.S. 1973. Transport of particles to surfaces, *Aerosol Sci.* 4:227–234.

19. Liu, B.Y.H., and J.K. Agarwal. 1974. Experimental Observation of aerosol deposition in turbulent flow, *Aerosol Sci.* 5: 145–155.

20. Sehmel, G.A., and S.L. Sutter. 1974. Particle deposition rates on a water surface as a function of particle diameter and air velocity, *J. Recherches Atmospheriques* 3:911–920.

21. Monteith, J.L. 1963. Dew, facts and fallacies, in *The Water Relations of Plants* (A.J. Rutter and F.H. Whitehead, eds.). John Wiley: New York.

22. Goldsmith, P., and F.G. May. 1966. Diffusiophoresis and thermophoresis in water vapor systems, in *Aerosol Science* (C.N. Davies, ed.). Academic Press: London.

23. MAP3S. 1981. The MAP3S/RAINE precipitation chemistry network: Statistical overview for the period 1976–1980, authored by The MAP3S/RAINE Research Community, submitted to *Atmos. Environ.*

24. Feely, H.W. and R.J. Larsen. 1979. The chemical composition of precipitation and dry atmospheric deposition. U.S. Department of Energy Environmental Quarterly Report EML-356, I-251 to I-350.

Measurement of Local Climatological and Air Pollution Factors Affecting Stone Decay

IVAR TOMBACH

The atmosphere is a primary contributor to the decay of stone in historic buildings. These atmospheric contributors range from the natural consequences of rainfall, wind, frost, and heat to the more complicated chemical and biological processes resulting from pollution. A list of such factors, though extensive, can be broken down into groups depending on: the available moisture (rain, fog, humidity); the temperature of the air; the cooling and heating of surfaces (by wind and radiation) and the evaporation and condensation of moisture on them; the motion of the air (wind); and the presence of air constituents and contaminants (gaseous and aerosol). The effectiveness of these factors depends on the time of day and seasons of the year, as well as on large-scale meteorological phenomena and human activities.

Techniques for measuring parameters within each group have been well developed in the fields of meteorology, aerodynamics, and air pollution. These methods can be applied to assist in research on stone preservation and can also provide data for developing strategies to protect specific structures.

Throughout the ages, stone has been used as a building material because it lasts longer than wood or other materials. Even the most permanent stone structures are subject to attack by nature, of course, but the typical time scale over which damage occurs from natural effects covers many human life spans (except when damage is caused

Ivar Tombach *is Vice President of Environmental Programs, AeroVironment, Inc., Pasadena, Calif.*

by cataclysmic events). However, human activities in an industrialized society have inadvertently contributed to a dramatic acceleration in the rate of decay of historic stone structures, to the point where the year-by-year decay of stone structures built decades, centuries, or even millenia ago is now often clearly perceptible.

Both the natural and human causes of such destruction of stone are becoming better understood, and efforts are being made throughout the world to preserve structures of particular historic significance. It is the purpose of this paper to aid in this preservation effort by evaluating some of the factors that cause stone decay from the viewpoint of atmospheric physics. The intent is to discuss ways to better understand the atmospheric conditions that influence the decay of a particular structure so that the preservationist can develop the best approach for protecting the structure. Such protective measures can range from control of external factors—say, by eliminating a source of decay-causing air pollution or by protecting a structure from rain or air pollution—to physical or chemical treatment of the stone itself.

ATMOSPHERIC VARIABLES AFFECTING STONE DECAY

The decay of stone can be caused by a variety of mechanisms.[1-7] These mechanisms can be classified into categories as shown in Table 1. Atmospheric factors that participate in these mechanisms are also shown in the table. An effort has been made to distinguish factors that contribute directly to a mechanism or to its destructiveness from those that participate more indirectly; the distinction is often subtle, and therefore the assignments are not necessarily unique. As an example of a secondary factor, the presence of atmospheric pollutants or aerosol is not necessary for changes to take place in the volume of material within interstices in the stone, but the material whose expansion causes the damage is often the by-product of an earlier chemical reaction with an atmospheric pollutant.

The mechanisms of stone decay require, almost universally, the presence of water (in either gaseous or liquid form), and many of the mechanisms require the existence of foreign materials in the stone or on its surface. These impurities are usually introduced to the stone by wet or dry deposition from the atmosphere, or are the by-products of chemical reactions with these atmospheric materials. The processes of wet and dry deposition of gases and particles are the subjects of another paper in this volume and therefore will not be discussed here.[8] Like the decay mechanisms, the deposition mechanisms depend on atmospheric factors, as summarized in Table 2.

The atmospheric factors that affect stone decay directly, and also affect the deposition rate, can be grouped into categories, as follows:

1. The available moisture (from precipitation, fog, humidity). Almost all decay mechanisms require some water, although heavy rainfall can wash away or dilute impurities and slow their attack on the stone. Hygroscopic aerosol particles grow at high humidities (typically, relative humidities greater than 70 percent) and are then more prone to gravitational settling or wind-caused impaction onto stone surfaces.

2. The temperature of the air. Damage occurs whenever the freezing point is crossed. Most chemical reactions proceed more rapidly as the temperature increases.

3. Solar insolation. Radiative cooling of stone at night can result in condensation of water on an otherwise dry surface and cooling or heating of the stone relative to the air affects deposition rates, as do evaporation and condensation.

4. Wind. The kinetic energy of abrasive particles and the degree of inertial impaction of particles or droplets onto the stone are dependent on the wind.

5. Air constituents and contaminants (gaseous and aerosol). Constituents in the air determine the rates of some forms of chemical attack and are often a necessary precursor of physical or chemical decay mechanisms. Natural constituents, such as CO_2 and sea-salt aerosol, play a role, as do manufactured pollutants. Obviously, the rate of deposition of a chemical is proportional to its concentration in the air.

To evaluate the significance of each of these factors in a given situation requires, first, an understanding of which mechanisms are potentially of concern for the type of stone, the construction method, and the foundation soil chemistry and moisture. By measuring the relevant atmospheric variables, it is then possible to determine, at least qualitatively, the potential diurnal and seasonal variability in the strength of the decay and deposition mechanisms. For example, Fassina has studied the effects of environmental conditions on the deterioration of stonework in Venice using daily measurements of meteorological conditions and of some atmospheric pollutants.[6]

In a few cases where a theoretical or empirical basis has been developed to describe a decay mechanism quantitatively, it may even be possible to predict the behavior and to compare the relative significance of several mechanisms. Chemical reaction rates are in this latter category,[9] along with the stresses caused by freezing water[10] and various

TABLE 1 Classification of Mechanisms Contributing to Stone Decay

Mechanism	Rainfall	Fog	Humidity	Temperature	Solar Insolation	Wind	Gaseous Pollutants	Aerosol
External Abrasion								
Erosion by wind-borne particles						●		●
Erosion by rainfall	●●							
Erosion by surface ice	●●	●		●				
Volume Change of Stone								
Differential expansion of mineral grains				●●			○	
Differential bulk expansion due to uneven heating				●●	●			
Differential bulk expansion due to uneven moisture content	●	●	●	●	●	○	○	○
Differential expansion of differing materials at joints				●				
Volume Change of Material in Capillaries and Interstices								
Freezing of water	●●	●●		●●				
Expansion of water when heated by sun	●	●		●	●			
Trapping of water under pressure when surface freezes	●	●		●				
Swelling of water-imbibing minerals by osmotic pressure	●		●					
Hydration of efflorescences, internal impurities, and stone constituents	●		●●				○○	○
Crystallization of salts		●		●	●	●	○	○○
Oxidation of materials into more voluminous forms	●	●					○	

Dissolution of Stone or Change of Chemical Form
  Dissolution in rainwater
  Dissolution by acids formed on stone by atmospheric gases or particles and water
  Reaction of stone with SO_2 to form water-soluble material
  Reaction of stone with acidic clay aerosol particles

Biological Activity
  Chemical attack by chelating, nitrifying, sulfur-reducing, or sulfur-oxidizing bacteria
  Erosion by symbiotic assemblages and higher plants that penetrate stone or produce damaging excretions

NOTE: Solid circles denote principal atmospheric factors; open circles denote secondary factors.

TABLE 2 Classification of Mechanisms of Deposition of Gases and Particles

Mechanism	Deposits of gases (G) or particles (P)	Relevant Atmospheric Factors					
		Rainfall	Fog	Humidity	Temperature	Solar Insolation	Wind
Dry Deposition							
Gravitational settling	P			○			
Inertial impaction	P			○			●
Brownian or molecular diffusion	G,P			○			
Stefan flow (toward surfaces where moisture is condensing)	G,P			●	●	●	
Thermophoresis (toward cold surfaces)	P				●	●	
Diffusiophoresis (toward evaporating surfaces)	P			○	●	●	○
Wet Deposition							
Precipitation	G,P	●					
Inertial impaction of fog droplets	G,P		●				●

NOTE: Solid circles denote principal atmosphere factors; open circles denote secondary factors.

deposition relations.[8] As an example of an empirical quantitative relationship, Hudec has developed regression expressions relating stone damage to quantifiable physical properties of the stone and to the degree of saturation of the stone and the freezing of internal water.[11]

MEASUREMENT OF ATMOSPHERIC FACTORS

The factors described above can be measured easily in some cases and with great difficulty, or not at all, in others. This discussion will briefly evaluate the availability of suitable measurement methods for these factors. The focus is on approaches that could be used by the stone preservationist, usually within the confines of a limited budget and without the aid of a meteorologist, atmospheric physicist, or air pollution specialist. The emphasis is on methods that can be used in an operational mode for long-term data gathering; additional techniques that are more exacting and labor intensive may be appropriate for specific short-term studies. Because the measurements discussed will often be unfamiliar to the stone preservationist, factors that should be considered in their use will also be mentioned.

Obviously, whenever appropriate meteorological or air pollution data are available from a government, university, or private monitoring station, the use of those data is the most efficient and least expensive way to acquire information. As an example, Winkler has studied meteorological effects on the deterioration of the National Bureau of Standards test wall using meteorological data from the Washington National Airport.[12] Such data may not always represent the meteorological conditions that are affecting a specific stone structure, however; some cases will be pointed out below.

A tabulation of measurement methods that might be useful for stone preservation work appears in Table 3. As a guide for acquiring suitable instruments, the purchase cost is described as "low" if the instrument costs less than $500, "moderate" if it costs between $500 and $2,000, and "high" if more than $2,000 is required. Operating costs are harder to quantify and depend considerably on the specific location of the study site. The same terms—"low," "moderate," and "high"—are used to describe operating costs, but only in a relative way. Similarly, the difficulty of using a given method (in terms of learning time, technician expertise, attention to detail, frequency of calibration, and difficulty of maintenance) is also indicated in relative terms using the same expressions. The comments below on the measurement methods supplement the material in the table.

TABLE 3 Selected Methods of Atmospheric Measurement for Stone Preservation Applications

Atmospheric Parameter	Method	Time Resolution	Automatic Recording	Purchase Cost	Operating Cost	Operating Difficulty	Comments
Rainfall	Manual rain gauge	Typically 1 h to 24 h	No	Low	Low	Low	Samples can be analyzed for rain chemistry.
	Recording rain guage	Minutes	Yes	Moderate	Low	Low	Collection of samples for rain chemistry has poorer time resolution.
Fog	Human observations	Typically a few hours	No	—	Low	Low	Does not quantify water content of air.
	Photoelectric sensors	Minutes	Yes	High	Moderate	Moderate to high	Does not quantify water content of air; instrument operation requires specialized advice.
Humidity	Sling psychrometer	Minutes	No	Low	Low	Low	Labor intensive.
	Hair hygrometer	Minutes	Yes	Low	Low	Low	Not as precise; poor accuracy above 90% relative humidity.
	Dew point sensor	Seconds	Yes	Moderate to high	Moderate	Moderate to high	Most precise; requires careful calibration.
Temperature	Thermistor or platinum resistance in radiation shield	Seconds	Yes	Moderate	Low	Low	Expert advice needed for use near walls or ground.
Solar insolation	Pyranometer	Seconds	Yes	Moderate	Low	Moderate	Difficult to use near walls.
	Net radiometer	Seconds	Yes	Moderate	Low	Moderate	Measures net radiation flux to/from surface; requires careful installation.

Wind	Hand-held wind gauge	Seconds	No	Low	Low	Low	Labor intensive.
	Cup and vane anemometer	Seconds	Yes	Moderate	Low	Low	
	Propeller anemometer	Seconds	Yes	Moderate	Low	Moderate	Can use near walls; can measure vertical airflow.
Gaseous air pollutants (SO_2, NO_x)	Sulfation plate (SO_2)	Days	No	Low	Low	Low	Semiquantitative; measures SO_2 deposition.
	Automatic analyzers (chemiluminescent for NO_x fluorescent for SO_2)	Seconds to minutes	Yes	High	Moderate	High	
Aerosol collection	Collection on filter—single broad range of particulate sizes	Hours	No	Low to moderate	Low	Low	
	Collection on multiple filters covering several narrower size ranges	Hours	No	Moderate to high	Low to moderate	Moderate	
Aerosol chemical analysis	X-ray spectroscopy (PIXE or XRF) of filter samples to determine elements Na and heavier	—	—	—	Ca. $25/sample	—	Requires complex equipment; analyses available from commercial labs.
	Wet chemistry (ion chromatography or colorimetry) of filter samples to determine $SO_4^=$, NO_3^-, or NH_4^+	—	—	—	Ca. $30–$50/sample	—	Requires complex equipment; technique also usable for rain sample analysis.
	In-situ S and Na particle monitoring by flame photometry	Seconds	Yes	High	High	High	Research techniques; instruments not commercially available.

Rainfall

For climatological purposes, daily rainfall data will suffice. For detailed studies, the intensity of rainfall (cm/min or cm/h) is relevant because more intense rainfall washes more effectively and also may be chemically less reactive. Data from a nearby government weather station may be sufficient, but rainfall can vary significantly over distances of a few kilometers. If the wetting of a specific wall is of interest, then a rain gauge has to be installed next to that wall. A windward wall will be wetted considerably more than a leeward wall. Architectural features can protect some portions of a wall.

Samples for pH measurement or chemical analysis can be those collected by the rain gauge, but it is frequently more practical to use a sample from a specifically designed collector. The samples are analyzed at an analytical laboratory using relatively standard techniques. Care has to be taken to avoid changing a sample's chemistry during collection and handling; it should not be kept in the sampler any longer than necessary, preferably no more than a day.

Fog

The parameter of greatest interest with fog is its liquid water content, which is measurable only with specialized research-grade samplers. The presence or absence of fog, and its visibility-impairing effects, can serve as useful indices of the presence of liquid water for many purposes, however. For qualitative purposes, visibility determinations at a nearby airport may be usable, but such information should be used cautiously since the existence of fog at a specific location depends considerably on the elevation or proximity to a body of water. The urban heat-island effect generally reduces fog in cities, but pollution from a city often increases fog downwind.

Humidity

Relative humidity is fairly easy to measure if great accuracy is not required. However, it is difficult (and expensive) to measure if, say, 1 percent accuracy is desired or when the humidity is near the saturation point of air. Local airport or weather service data may be adequate for many situations. In this case one should use dew point, rather than relative humidity, because dew point is a characteristic of the water content of the larger-scale air mass, while relative humidity depends on the local temperature and therefore depends on local factors. Dew point is converted to relative humidity using the temperature measured at the study site.

Temperature

Proper measurement of air temperature requires that the sensor not be cooled or heated by radiation and hence that it be installed in a radiation shield. Expert advice should be sought for selecting the appropriate shield for use near a wall to avoid reflected radiation. Because of local heating and limited air circulation, the air temperature near the walls of a building could vary from one side of the building to the other.

Thermistors can also be embedded in the stone to measure the wall temperature. The most useful measurement location is probably as close as possible to the exterior surface. Circuits are available, at reasonable prices, that can compare the temperatures measured by the two matched sensors in the wall and in the air with an accuracy of better than 0.1° C.

Solar Insolation

Solar insolation is a guide to how much the sun's radiation contributes to temperature changes in a wall. Because local shadows affect the extent of solar insolation, measurements should be made as close as possible to the portion of the wall that is of interest, unless only a general characterization of the amount of insolation is needed.

Standard meteorological sensors of solar insolation can be used to measure insolation on a wall if they are oriented parallel to the wall's surface. Similarly, net radiometers are available to measure both the solar radiation incident on the wall and the radiation emitted by the wall.

Because solar insolation measurements for stone preservation research are somewhat unusual, expert advice should be sought on the appropriate sensor, its installation, and the interpretation of data from it.

Wind

The low-speed end of the wind spectrum is of interest for diffusion and the high-speed end for abrasion. Most inexpensive wind sensors lack a sufficiently low starting threshold to cover the low speeds characteristic of early morning hours.

The direction of the wind may not be of interest if near-wall measurements are being made; the airflow is necessarily parallel to the wall there (but frequently has a vertical component). For such work, propeller anemometers are more useful than the more conventional cup-and-vane sensors.

Because the wind depends so much on local obstructions, weather service or airport data are useful only as indicators of general direction and speed of the airflow through a region.

Gaseous Air Pollutants

Concentrations of air pollutants in and near cities vary dramatically from hour to hour (or even from minute to minute). Therefore, only an instrument that can respond to these changes can lead to meaningful assessments of the effects of pollutant fluxes to material surfaces. The same sort of response is needed from the sensors that determine whether there is a deposition flux toward the surface at any given time. Thus, sensors that integrate pollutant concentrations over long periods are generally not useful for detailed studies because the average deposition flux of material to stone surfaces depends not only on the average concentration, but also on the correlation of the concentrations with a positive deposition flux. Sulfation plates are an exception. They are long-term sensors that are potentially useful because they directly measure the deposition of SO_2 to the plate.

Ideally, one would like an existing air pollution monitoring station within a kilometer or two of the study site, with no nearby pollution sources to cause the concentrations of SO_2 or NO_x at the study site to differ from those at the station. Otherwise, air pollutant monitoring becomes an expensive venture, and expert help will certainly be needed to calibrate the instruments. Fortunately, commercially available instruments (especially those certified as "reference or equivalent methods" by the U.S. Environmental Protection Agency) are stable and reliable when properly used. Fully self-contained dry methods exist for detecting both NO_x and SO_2; there is no need to deal with the complexity of sensors that require auxiliary compressed gases or perform analyses by wet chemistry. Although the state of the art of air monitoring is changing rapidly, Stern provides an excellent starting basis for understanding the science.[13]

Aerosols

Automated aerosol analyzers that would be appropriate for stone-preservation research do not exist (with one exception mentioned below).

The primary interest is the chemical content of the aerosol, especially the presence of salt (NaCl), sulfates ($SO_4^=$), nitrates (NO_3^-), and the ammonium ion (NH_4^+). The usual technique, therefore, is to draw air through one or more filters and analyze the collected material in the laboratory. Lundgren et al. provide a useful reference on the state of the art of aerosol collection and analysis.[14] Stern also covers the subject, but in a more introductory manner.

The filter material is critical because some particles and gases react with the filter and form "filter artifacts" and because the filter has to be compatible with the analysis procedure. For most purposes, Teflon or Teflon-coated filters are the most appropriate. Polycarbonate filters (Nuclepore) are necessary for electron microscope analyses; nylon filters collect gaseous nitric acid (HNO_3) in the air; and prefired quartz filters are required if an analysis for carbon or soot is planned.

The filters can be analyzed at a commercial, university, or government laboratory that is familiar with the handling of air pollution samples. X-ray spectroscopy techniques—PIXE (particle-induced X-ray emission spectroscopy) and XRF (X-ray fluorescence)—are inexpensive and describe much of the composition of the aerosol. Such techniques directly identify the NaCl content; they also indirectly provide the $SO_4^=$ content, and experience has shown that essentially all the sulfur in the air is in the sulfate form. Wet chemical techniques are needed to identify NO_3^- and NH_4^+ and are also appropriate for $SO_4^=$.

The two-stage approach—sampling followed by analysis—need not be followed to assess sodium and sulfur-containing particles. In situ measurement of these elements (and thus, for all practical purposes, of NaCl and $SO_4^=$) in particles has been performed by modifying commercially available flame-photometric air pollution analyzers. Pueschel[15] describes a sodium particle analyzer, and Coburn et al.[16] and Huntzicker et al.[17] describe sulfur particle analyzers. Although these techniques are not available off the shelf, they have sufficient utility in some research applications to justify the special expertise needed to use them.

CONCLUSIONS

The atmosphere exerts a significant influence on both natural and human-related mechanisms of stone decay. Although a complete quantitative theory of stone decay is unlikely because of the many site-specific variables, it is possible to infer relationships between stone decay and atmospheric conditions. In most cases relatively standard instruments used by meteorologists and air pollution scientists can be applied, with perhaps minor adaptations, to studies of stone decay. Stone preservationists therefore do not need to develop methods for atmospheric measurements; they are also able to draw on the expertise of the meteorological and air pollution scientific communities for assistance in their efforts.

REFERENCES

1. Torraca, G. (1976) Brick, adobe, stone, and architectural ceramics: Deterioration processes and conservation practices. *Preservation and Conservation Principles and Practices*. Preservation Press, National Trust for Historic Preservation in the United States, Washington, D.C., 143–165. Vittori, O. (1976) Secondary sinks in atmospheric gas dispersion models. Proceedings Seminar on Air Pollution Modelling, IBM Italy, Venice Scientific Center, 27–28 November 1975.

2. Keller, W.D. (1977) Progress and problems in rock weathering related to stone decay. *Engineering Geology Case Histories Number 11*, Geological Soc. of Am., 37–46.

3. Winkler, E.M. (1977) Stone decay in urban atmospheres. *Engineering Geology Case Histories, Number 11*. Geological Soc. of Am., 53–58.

4. Hyvert, G. (1977) Weathering and restoration of Borobudur Temple, Indonesia. *Engineering Geology Case Histories Number 11*, Geological Soc. of Am., 95–100.

5. Gauri, K.L. (1978) The preservation of stone. *Scientific American*, June.

6. Fassina, V. (1978) A survey on air pollution and deterioration of stonework in Venice. *Atmos. Env. 12*, 2705–2211.

7. Hansen, J. (1980) Ailing treasures. *Science 80, 1*, 58–110.

8. Hicks, B.B. (1981) Wet and dry surface deposition of air pollutants and their modeling. National Academy of Sciences Conference on the Conservation of Historic Stone Buildings and Monuments, Washington, D.C., 2–4 February, 1981.

9. Vittori, O. (1976) Secondary sinks in atmospheric gas dispersion models. Proceedings Seminar on Air Pollution Modelling, IBM Italy, Venice Scientific Center, 27–28 November 1975.

10. Winkler, E.M. (1973) *Stone Properties Durability in Man's Environment*. Springer-Verlag, New York–Vienna, 250.

11. Hudec, P.P. (1977) Rock weathering on the molecular level. *Engineering Geology Case Histories Number 11*. Geological Soc. of Am., 47–51.

12. Winkler, E.M. (1981) Problems in the deterioration of stone. National Academy of Sciences Conference on the Conservation of Historic Stone Buildings and Monuments, Washington, D.C., 2–4 February, 1981.

13. Stern, A.C., ed. (1976) *Air Pollution*, Vol. III, 3rd edition. Academic Press, New York, 797.

14. Lundgren, D.A., F.S. Harris, W.H. Marlow, M. Lippman, W.E. Clark, and M.D. Durham, eds. (1979) *Aerosol Measurement*. University Presses of Florida, Gainesville, 716.

15. Pueschel, R.F. (1969) Thermal decomposition of sodium-containing particles in a flame. *J. Colloid and Interface Sci., 30*, 120–127.

16. Coburn, W.G., R.B. Husar, and J.D. Husar (1978) Continuous *in situ* monitoring of ambient particular sulfur using flame photometry and thermal analysis. *Atmos. Env. 12*, 89–98.

17. Huntzicker, J.J., R.S. Hoffman, and C.S. Ling (1978) Continuous measurement and speciation of sulfur-containing aerosols by flame photometry. *Atmos. Env. 12*, 83–88.

Diagnosis of Nonstructural Problems in Historic Masonry Buildings

BAIRD M. SMITH

This paper explores the deterioration of stone and brick in buildings and monuments caused by defects in building design or by materials that have proven nonfunctional or of lower than expected performance and durability. For instance, frequent problems occur from the corrosion of metal anchorage devices, improper design of stone details precluding proper drainage of rainwater, and the unexpected ramifications of newly installed air conditioning or the introduction of thermal insulation. A variety of cases are identified to aid in the proper diagnosis of problems and determination of treatments. The paper does not deal with deterioration from primary sources, such as airborne pollutants, freeze/thaw cycling, or thermal stressing, but rather with deterioration resulting from improper construction practices, weak design details, or the use of materials that have since proven unsatisfactory.

This report is different from most of the others in that it does not deal with deterioration associated with airborne pollution, rain, and other moisture-related problems, but with deterioration from secondary causes. For example, the report explores causes such as poor craftsmanship or construction practice, or materials and building systems that simply have not performed as well as expected. Thus, topics such as freeze/thaw, acid attack, and so on are not mentioned; the focus is on building

Baird M. Smith *is Historical Architect, Heritage Conservation and Recreation Service, U.S. Department of the Interior, Washington, D.C.*

practices, with the emphasis on buildings and techniques used from the late nineteenth century to the present.

The approach here is to look at the basic physical elements of a building, then to describe the evolution of the building technologies affecting those elements. Thus, after presenting some overriding conditions, the following building elements are discussed:

- Walls
- Top of Walls
- Window Openings
- Base of Walls

In each case, emphasis is placed on an evaluation of actual building practices to determine causes for common problems. Proper identification and understanding of the earlier building systems is indispensable when attempting to diagnose current problems and prescribe treatments. This report should aid architects, engineers, and builders in the proper diagnosis of such problems.

Throughout this report, some rather extreme cases of building failure are cited. This is not meant to overdramatize the topic, but rather to present a reasonably accurate assessment of the potential for modern building failures resulting from poor original building practices.

One must remember that from 1880 to 1940 there was tremendous competition among builders, architects, and product manufacturers. Many systems and products were proprietary; hence, there was an attempt at secrecy. Also, many products were used without proper testing, not to mention poor craftmanship caused by owners and builders cutting corners with construction budgets and timetables. Most systems and products were developed through trial and error. It was an age of exploitation of building materials and systems. Today we are left with that legacy. Unfortunately, the preservation of some of these buildings may be financially impossible. Of course, other early buildings and systems show little potential for failure or severe deterioration; therefore, little should impede their successful preservation.

Research for this paper was based on published literature of the period. Unfortunately, few records of building failure were kept, and rarely were ineffective building products or systems criticized by writers of the time. Thus the findings of this report relate to the comparison and evaluation of known building and product failures. Clearly, more research should be done to support further many of the deductions presented here.

OVERRIDING CONDITIONS

Several aspects of good building practice that are generally known today were rarely understood previously. Major examples are described here.

First, it has long been recognized that when choosing a building stone or brick, one should match the expected weathering performance of the material to its expected use or exposure. Thus, stone at the top of a wall or at the cornice must be more durable and more capable of withstanding severe weather than stone on the lower part of the wall. However, in the competitive building market this was not generally understood; hence, the most durable materials often were not selected for use where they were most needed. Today; deterioration of such weak stones or poor quality brick is common. The materials were simply used incorrectly.

Second, ornately carved stone may not be practical where exposed to extreme weather. There is an incongruity here in that soft stone, which is easiest to carve, is generally the least durable. Thus, a heavily carved balustrade at the top of the wall is the worst possible location for soft stone. Unfortunately, this point was often ignored, so we are now faced with treating severe deterioration of stone or brick that has little hope of lasting very far into the future.

Third, there are many cases of chemical and physical incompatibilities among building materials. Some examples such as acids with limestone or salts within masonry are obvious. But other examples are less obvious, as in the following:

- Portland limestone and brick. Rainwater washing down limestone onto brick will result in stains and surface damage to the brick. There is little that can be done to arrest this process.
- Portland cement and calcareous stones. Ferrous oxide in portland cement will always stain limestone, marble, and some sandstones. Nonstaining portland cement has been developed to eliminate this problem.
- Iron, steel, and copper with masonry. Oxidation of these metals can cause both staining and physical damage to stone and brick. Careful attention must be paid if these materials come in contact with masonry.

Fourth, overall building profiles have changed. A comparison of a wall section from an eighteenth century stone building and one from the mid-twentieth century would clearly show the evolution away from control of rainwater on the face of a building. The early building would have a projecting cornice, pediments over each window, projecting belt courses, and a water table. All these features were intended to direct rainwater off the face of the wall, minimizing staining, weathering, and other moisture problems. In contrast, facades of recent buildings are comparatively flat, with little or no attention to the control of rainwater. Thus, moisture problems can be common.

WALLS

The most basic element of a building is its walls. For the purpose of this report it is shown that walls have evolved through time from simple bearing walls, to "cavity walls" (bearing walls with a vertical air cavity), to nonbearing "curtain" walls. In most instances, nonbearing walls replaced bearing walls (with the exception of residential-scale buildings) beginning in the 1880s and were in common use by the turn of the century. This is one of the most interesting evolutions in terms of changing technologies, introduction of new building materials, adaption of traditional designs, and new engineering approaches.

Bearing Walls

To diagnose problems in masonry bearing walls properly one must understand the typical building practices employed. Cut stones of from 10 in. to more than 36 in. wide were laid one on top of the other, always with their bedding planes in their natural (horizontal) position. Brick walls were built up of two, three, or four wythes solidly packed with mortar. The mortar was normally soft, since it was made principally of lime. Its purpose was to form a uniform bearing plane between masonry units—in effect, to hold the units apart. It did not hold the units together—that is, it did not create adhesion.

Problems can now occur with these simple wall constructions where individual stones are connected with dowels, anchors, or cramps (see Figure 1). Such devices were often used with projecting stones (at a cornice or over a window) to provide a sufficient tie or connection to the main bulk of masonry. Before the widespread availability of steel (by the late nineteenth century), these metal anchors were generally made of wrought iron. They were anchored into the stone with molten lead, a sulfur and sand mix, or a mortar grout. The wrought iron generally provided resistance to severe corrosion as long as it was

FIGURE 1 Early anchorage devices. Two types of cramps (or clamps) are shown at the left, and a dowel is at the right. These were made of wrought iron, sometimes in cast iron, later in steel, and occasionally in bronze or copper.

sufficiently buried within the wall to remain dry. Severe corrosion would occur upon exposure of the anchors to the weather, often jeopardizing the integrity of the connection. This could result in a failure or collapse of the stonework affected.

It is now known that wrought iron, cast iron, or steel anchors set in sulfur or in mortar can become severely corroded if moisture reacts with the sulfur (forming sulfuric acid) or with the mortar (forming carbonic acid). If such conditions exist, one should attempt to investigate all connections to determine the degree of deterioration. Unfortunately, thorough investigation is rarely feasible, and drastic measures such as complete removal of masonry may have to be considered. However, this is rare, and in the universe of potential nonstructural problems in masonry construction, bearing walls are the least problematic.

Cavity Bearing Walls

By the early nineteenth century, it was widely recognized that walls of solid masonry were susceptible to moisture penetration from hard-driving rain. To prevent this penetration of moisture from the outside, a vertical air space or cavity (of perhaps 2 to 4 in.) was introduced into the center of the wall (see Figure 2). Thus, two wall sections—the outer of dressed stone or hard-fired bricks and an inner of backfill rubble stone or irregular brick—were laid independently, yet simultaneously. Generally both were bearing the weight of the wall above, except in very thick walls, perhaps more than 24 in., where the weight was carried on the inner wall.

The cavity was designed to provide an outlet for the moisture that penetrated the outer wall. The moisture would flow to the base of the

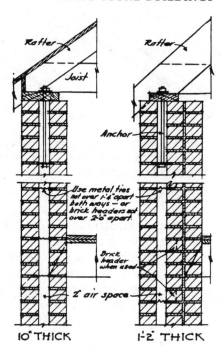

FIGURE 2 Typical brick bearing wall construction. This detail for common residential scale buildings shows the wall cavity with corrugated metal ties (probably of galvanized steel). SOURCE: *Architectural Graphic Standards*, 1936, p. 16.

cavity and through weep holes or other small openings toward the outside (either toward an interior crawl space or to the exterior at the base of the wall).

The wall sections were positively tied together, either with various types of metal ties or with bond stones, bricks, or structural terra-cotta turned at right angles to the face of the wall and bridging across the cavity to the inner wall section. The metal ties were first made of wrought and cast iron; then, by the mid-nineteenth century, galvanized iron, painted iron, iron dipped in tar, and even copper or bronze were used. Early literature shows that some of these cavity-wall ties were shaped to reduce corrosion problems by eliminating flat surfaces (potential catch basins for moisture in the cavity). Figure 3 illustrates some of the variations.

There are a host of typical problems with cavity walls. First, there can be differential settlement between the inner and outer wall sections, weakening or even breaking ties between the two and thus jeopardizing the structural integrity of the wall. Second, during construction, the cavity can become filled with mortar or debris, clogging weep holes or resting on top of the metal ties and thus creating a small

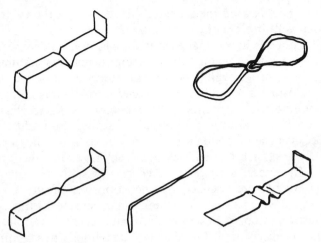

FIGURE 3 Typical metal cavity-wall ties. These ties could be of galvanized or painted iron and steel, copper, bronze, or even plastic. They are shaped with the special twists or corrugations to prevent water from standing on any flat surfaces.

catch basin for moisture. Third, the metal anchors (especially painted ones or those of galvanized iron) are susceptible to corrosion because the air cavity is often very moist. Therefore, through time these metal ties have often failed because of corrosion, which results in a loss of the wall's structural integrity.

Cavity-wall construction can be identified through physical probing, which involves careful core drilling or discreet dismantling of a portion of a wall, or with various nondestructive techniques. Naturally, any original architectural or engineering drawings and/or building specifications are invaluable in the identification and diagnosis procedures.

Nonbearing or Curtain Walls

Probably the most interesting topic area in this report is the development of the curtain walls and skeleton steel and concrete construction of the first part of the twentieth century. Such walls were rarely thicker than 12 in., and their weight was carried either on each floor or on every other floor, supported by the floor framing. Obviously, there were many changes in the technology of construction and types of building materials used. Elevators, skyscrapers, fireproof construction, and lightweight building skins were some of the major elements and new technologies developed during this period. To explore those developments is beyond the scope of this report, but the evolution of the curtain wall itself is of central interest.

The first issue is in the change of materials. Bricks or stones, which are quite heavy and capable of supporting themselves, naturally had little use in a curtain wall. The key function was to provide a lightweight enclosure, albeit a decorative one. Strength was not an issue. Thus, sandlime bricks, architectural terra-cotta, cast stone, glass, metal sandwich panels, and so forth were all substitutes for traditional masonry. Because they were lightweight and nonbearing, they had to be extensively tied back to their masonry backing and to the structural supports. The types of anchors varied greatly, both in shape and in materials, but all were intended to be noncorrosive. The most common materials used were painted or galvanized iron and steel wires. In some cases, bronze or copper ties were used, and recently, various plastic materials have been tried. The shapes varied, from wires (a minimum of #8 gauge or about 1/8 in. in diameter) tied to the masonry unit and to the backing (see Figure 4), to shaped or bent pieces, or to sheets of wire cloth. With either brick or terra-cotta, these wires or ties were needed continuously throughout the wall, often spaced as closely as 8 in. centers both vertically and horizonally.

In the early part of this century, painted steel was considered noncorrosive. This was found to be untrue—after a time it corroded badly—so galvanized steel became the minimum required. Stainless steel was introduced after World War II, but galvanized and painted iron never fell out of favor. They may still be used today on "budget" jobs, but they are not considered to have anywhere near the life expectancy of stainless steel. Note, however, that some of the early stainless steels have proven to be corrosive. One must be careful when selecting from several grades available today to assure that an anchorage device capable of withstanding the corrosive conditions within a wall is chosen. Generally, types 301 and 302 are satisfactory today.

An ancillary problem with materials is that the facing stone on curtain walls is often only 2 in. thick and generally face bedded rather than bedded in its natural plane. The result can be delamination and extensive surface spalling. Stone with the potential for delamination must be laid in its natural bedding plans.

A second issue is the wall cavity. There was no question that a cavity was efficient at preventing moisture penetration from the rain. The evolution and treatment of the cavity in curtain-wall construction is interesting. In early curtain walls, the cavity was treated just as it had been in bearing walls. That is, there was an outer veneer of stone, brick, or terra-cotta (generally 4 in. thick), then the air cavity (from 2

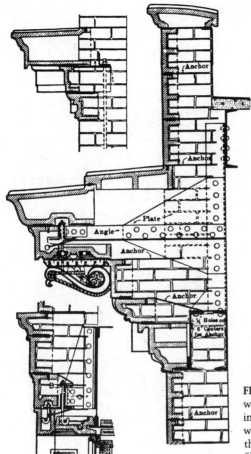

FIGURE 4 Anchorage techniques with terra-cotta. Each piece is individually tied back with metal wires to the backup masonry and the structure. SOURCE: *Masonry: A Short Textbook*, 1915, p. 86.

to 4 in. thick), and then the inner masonry backup material resulting in a total wall thickness of about 12 in.

With this type of wall, new details had to be devised to sheath the structural iron with masonry. Figure 5 is an example. Note that the structural iron was encased with stone, that the resulting cavity was vented to the outside (note words "open joint"), and that the stone was anchored with metal ties. Although the encased iron and steel was painted, often with up to three coats of iron oxide paint, corrosion was still common. To correct this problem, the cavity was eliminated and all steel or iron was thoroughly covered or encased with mortar grout. However, the resulting solid masonry wall could suffer moisture

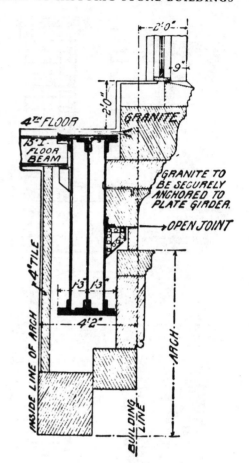

FIGURE 5 Stone encasement for floor beams. The stone is a veneer, with a sizeable inner wall cavity to prevent rain penetration. However, rain penetration was still found to be a problem, and corrosion to the steel was common. SOURCE: *Architectural Engineering*, 1901, p. 154.

penetration, causing disruption to the interior wall finishes. Therefore, the inner plaster was not placed directly on the masonry, but rather on furring strips, resulting in a 1/2 in. air cavity. This is often known as a "furred" wall and is now the preferred detail.

The third issue is related to problems with anchorage and to changes in the cavity wall. It concerns the addition of flashing within the wall cavity at selected points to protect critical connections and anchors. Figure 6 illustrates the best in building practice in this regard after World War II. Note particularly the flashing at the spandrel beam, which protects the steel shelf angle at the window from moisture flowing down the cavity. Since this arrangement represents the best in practice, it can be assumed that many recent buildings would not

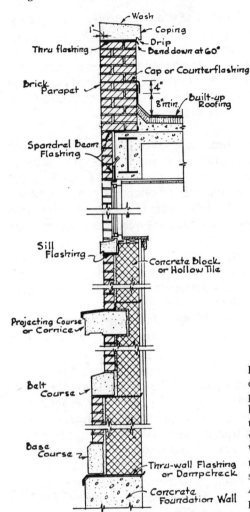

Wash
Coping
Drip
Bend down at 60°
Thru flashing
Cap or Counterflashing
Brick Parapet
4"
8"min. Built-up Roofing
Spandrel Beam Flashing
Sill Flashing
Concrete Block or Hollow Tile
Projecting Course or Cornice
Belt Course
Base Course
Thru-wall Flashing or Dampcheck
Concrete Foundation Wall

FIGURE 6 Wall flashing. Note the extensive use of wall flashings, especially at the spandrel beam and below the window sills. This practice is comparatively recent and would rarely be found before World War II. Note also that the inner plaster is furred out from the wall. SOURCE: *Materials and Methods of Architectural Construction*, 1964, p. 213.

have this type of flashing and, therefore, that there are some critical areas not adequately protected.

The fourth issue is the mortar itself. Lime mortar expands and contracts with changes of temperature or moisture content. In a bearing wall, the mortar joint is strictly in compression, and the expansion or contraction of materials has little effect on the structural integrity of the wall. With the evolution of the curtain wall, builders changed to a stronger, waterproof, portland cement mortar, usually a mixture of one part cement to one part sand. The precise reason for the change is not clear. There was, however, a reference to lime mortar's failing in a fire, and there was the need to make the joint as waterproof as possible to protect the anchorage devices and structural framework (portland cement mortar is nearly impermeable). The result recognized today is that these high strength cement mortars fail to prevent rain penetration. There are three obvious reasons for this:

• Portland cement mortars shrink upon setting. Since a curtain wall is not strictly in compresson, minute horizontal fractures from shrinkage occur. This has long been recognized. Often an elastomeric joint sealant or caulking was used in selected horizontal joints around a building, generally at the top of each section of the curtain wall where it joins a floor or spandrel beam. Through the years these joints open up, creating an entry point for rain.

• In tall buildings, lateral wind loads have the effect of opening up joints on the windward side of a building, placing tensile forces on the curtain wall. Obviously this breaks the mortar bond with the masonry, creating thousands of entry points for rain.

• Portland cement reacts with airborne sulfates, resulting in "sulfate attack." This causes expansion of the mortar, disrupting the mortar bond with the masonry.

These problems taken together create the potential for severe deterioration of curtain wall materials, especially iron and steel anchorage and structural systems. There may be some extant buildings suffering the worst of these problems that are time bombs waiting to go off. Left unattended, such deterioration could result in severe damage to individual building elements or, in extreme cases, in a failure of a structural component. To counter these potential problems, modern high-rise buildings rely heavily on flexible joint sealants. If the best materials or techniques are not used, the results can be disastrous.

TOP OF THE WALL—PARAPETS AND CORNICES

Parapets

A parapet is one of the architectural elements perhaps most susceptible to deterioration and damage. Because it is freestanding and not warmed by interior building heat, it is exposed to the worst weather and is subjected to harsh freeze/thaw cycling.

Builders have recognized the vulnerability of parapets and have gone to great lengths to provide protection to parapet walls. Such a wall must be topped with a capstone (also called a coping). Figure 7 illustrates the best practice with a capstone: The top is sloped to provide a water wash, and the stone extends beyond the face of the wall, with a reglet or drip on the underside. Where one capstone is joined to another, the mortar joint is very susceptible to water damage because it faces up. Hence, it is often packed with lead or tar or is actually flashed with galvanized iron or copper. This explains the best building practice regarding capstones, and little has changed from medieval times to the present.

There can be problems, however, even with capstones like the one in Figure 7. The need for flexible joints to accept horizontal thermal expansion in the parapet wall does not appear to have been recognized until early in this century. Without an expansion joint, the capstone can become dislodged, creating vertical points for water to enter the stone, brick, or terra-cotta parapet wall below. Now, expansion joints are recommended every 20 linear feet.

Cornices and Overhangs

In a building of simple bearing-wall construction, a stone cornice would merely be corbelled or cantilevered out beyond the stone below. Sometimes these cornice stones would have to be tied back to the main

FIGURE 7 Typical capstone. The sloping surface (a) is called a wash and assures that water will not stand on the top surface. The "reglet" or "drip" (b) prevents water from flowing under the capstone and down the wall. SOURCE: *Elements of Brick and Stone Masonry*, 1930, p. 39.

wall with wrought iron cramps. The cramps might be inadvertently exposed to the weather and require ongoing surveillance and maintenance to avoid serious corrosion and damage to the stonework.

As buildings became taller, and architects wanted to preserve some of the proportional relationships between the base, shaft, and capital or cornice of a building, the cornices became massive overhangs of stone or terra-cotta, suspended from the structure's frame. Figure 8 illustrates one such cornice, an extreme example of poor design and naiveté in understanding the potential for problems. The uppermost horizontal surface would have been roofed, probably with a form of composition roofing and gravel, with collected rainwater draining into interior leaders or downspouts.

The potential for failure of this building assembly is great. The bolt heads and nuts under the soffit are exposed to the weather, and other steel angles are very close to the surface of the stone (at points a and b) and susceptible to corrosion. Finally, composition roofing rarely lasts

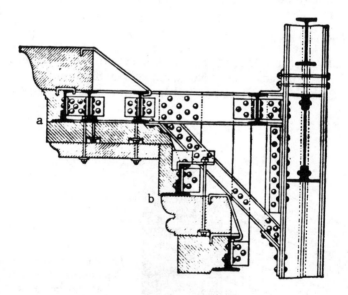

FIGURE 8 Projecting stone cornice. This stone cornice was naively designed. The bolt heads under the soffit are actually exposed to the weather and several angle irons (a and b) are perilously close to the outside face of the stone. SOURCE: *Architectural Engineering*, 1901, p. 176.

beyond 30 years without some cracks and leakage occurring. For water to enter this assembly would be disastrous. Although this is an extreme example, there are more than a few recorded cases where a piece of a stone or terra-cotta cornice has dropped to the ground.

A related development in cornice design is illustrated in Figure 9. Rather than sloping the top surface of the cornice stone outward, it was sloped inward to channel the collected water into the roof's drainage system. This change was brought about because the former design caused water to run off the cornice, thus staining portions of the wall below. Unfortunately, although the new design may have reduced staining, it greatly increased the potential for deterioration of the flashings and damage to the cornice and parapet because the rainwater is collected, rather than allowed to run off. Wherever water collects, the potential for damage is greatly increased. This new design for a cornice is not an improvement; it is a potential contributor to wall damage.

Roof Drainage

A not-so-subtle change in roof drainage has also occurred through time. The most common early technique was to pitch the roof (hip, gable, or gambrel roofs) and direct rainwater off and away from the building. Later, gutters and downspouts were added to control the water and assure that it was properly transported down and away from the base of the wall.

Eventually, as buildings got taller, the downspouts (most often of cast iron) had to be built into the wall. Two potential problems were created by this approach. First, if the downspouts were too close to the outside surface of the wall, they could freeze in winter, possibly fracturing the cast iron. Damage to the wall would result. Second, if the downspouts became clogged with leaves or other debris, it was generally impossible to clean them out because they were encased in masonry. Again, they could leak, causing wall damage.

Gutters can also be a source of problems with masonry. In many cases, they are incorporated into a stone or terra-cotta cornice and become hidden gutters. Figure 10 illustrates such a typical gutter in a stone cornice. The gutter could be lined with either copper or lead, but should the metals corrode or crack, or the joints open up, water will enter the cornice and top of the wall and cause damage. This type of gutter must be periodically inspected to assure that leaves or debris do not block proper drainage.

STONE

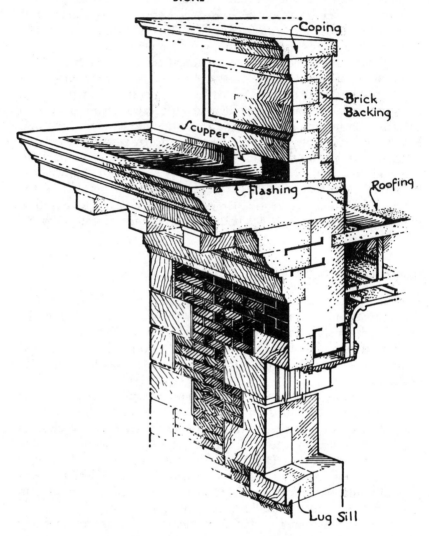

FIGURE 9 Modern stone cornice. This cornice design directs rainwater toward the roof and a central water control system. The design requires extensive flashing, which is especially subject to corrosion and damage to the surface. SOURCE: *Materials and Methods of Architectural Construction*, 1964, p. 122.

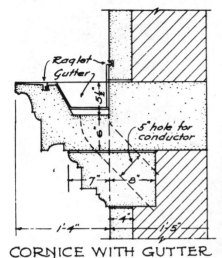

Raglet
Gutter
5½"
6"
5" hole for conductor
7"   8"
1'-4"   1'-5"

CORNICE WITH GUTTER

FIGURE 10 "Hidden" gutter in a stone cornice. The metal gutter lining, usually of copper or lead, is very susceptible to corrosion and damage. Without periodic maintenance, gutters can fill with leaves or debris, trapping water that can then corrode the metal lining and cause water to seep into the stone and wall assemblies below. SOURCE: *Architectural Graphic Standards*, 1936, p. 34.

WINDOW OPENINGS IN MASONRY

Sills

A common location for masonry deterioration or damage is at window (and door) openings. Because these openings penetrate wall enclosures, moisture may enter the center portions of the wall. Generally, the damage occurs at the head and sill of the window. Sills must be sloped outward and undercut with a drip or reglet, as shown in Figure 11. This sill is an example of the best practice and is known as a "lug" sill. It is distinguished from a "slip" sill by the fact that its ends are set a few inches into the masonry (see a), whereas the slip sill is shortened and is set within the masonry opening.

Slip sills were intended for "economical" construction where long building life was not expected. Figure 12 illustrates the points most vulnerable to water damage and also illustrates the staining pattern that often results. Any sill must project beyond the face of the wall below, or staining and deterioration will result there.

Because sills are exposed to weather on top and on the front, they are susceptible to deterioration, especially from freeze/thaw cycles. If they are face bedded, serious delamination and spalling can result.

Lintels

The head or lintel above a window opening is another common problem area. Traditionally, lintels were of wood or masonry. Masonry

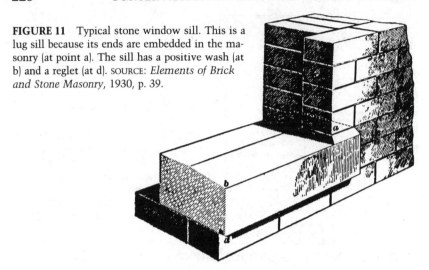

FIGURE 11 Typical stone window sill. This is a lug sill because its ends are embedded in the masonry (at point a). The sill has a positive wash (at b) and a reglet (at d). SOURCE: *Elements of Brick and Stone Masonry*, 1930, p. 39.

lintels were either a single stone spanning the full width of the opening or a combination of stones or bricks forming an arch. Arches could be flat, segmental, or pointed, but they all carried the weight of the masonry wall above.

With the availability of iron and steel angles, masonry arches were eliminated. Figure 13 illustrates a typical installation. Because the angles are partially exposed to the weather, the potential for corrosion is great. Little provision is made for rainwater coming down the face

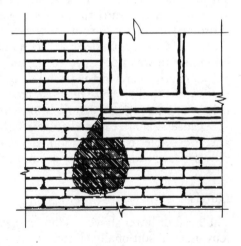

FIGURE 12 Staining from "slip" sills. Water flowing off the sill invariably stains the wall and seriously erodes mortar joints in the damage area (shown with cross hatching).

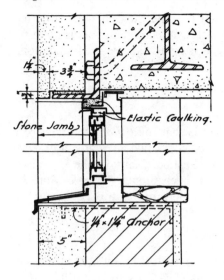

FIGURE 13 Steel angle at a window opening. The steel angle must remain painted (both within the cavity and on its bottom face) or corrosion and perhaps failure will result. Notice the absence of wall flashing. SOURCE: *Architectural Graphic Standards*, 1936, p. 32.

of the wall above and under the lintel. The water would just flow back to the angle, thus increasing corrosion. It is not uncommon when investigating a building to find steel angles at the lintel heavily corroded, with enough rust built up to disrupt the masonry above and loosen the anchorage of the angle. In such cases very little can be done; they represent serious repair problems.

BASE OF WALLS

Problems associated with groundwater, surface water, and related salt or freeze/thaw problems have been reasonably well understood since the mid-nineteenth century. Building practices designed to avoid such problems have remained largely unchanged during this century.

Problems beyond the obvious in older masonry foundations generally relate to a breakdown in materials or systems intended to keep water out of basements. For instance, outside surfaces of foundation walls were often coated with cement pargeting or asphaltic mastics. Through the years, these coatings become dislodged and no longer block moisture penetration. Damage to the foundations, especially to the mortar, results.

For most buildings built on damp soil, footing or foundation drains were quite common, especially during the twentieth century. These drains of clay tile or masonry often become clogged with silt, tree roots, or other organic or animal matter, thus reducing or eliminating drainage of groundwater. Naturally, this could damage the foundations and jeopardize building stability.

CONCLUSION

The obvious conclusion of this paper is that there are potentially very serious problems that have yet to be identified in many buildings. Early building practices were often slipshod. Many of the materials, techniques, or methods used had not been properly tested. More research is needed in several of the areas identified in this paper, but the issues are perhaps the following:

- There is a need for in-depth research into building practices between the 1880s and the present to be able more fully to understand stone and brick detailing, anchorage and flashing techniques, and assembly methods. Such understanding would aid in choosing appropriate repair and preservation treatments.
- To further understand early building techniques, more effort should be put into documenting building failures and into recording buildings being demolished or dismantled. A great deal of information about deterioration and damage to building elements could be acquired in this way.
- There are many problems with early high-rise buildings. One in particular is the repair of deteriorated mortar joints in masonry curtain walls. In attempting to reestablish a waterproof joint, should traditional "tuck pointing" techniques with a waterproof mortar be used, or should the joints be wiped with a "slurry coat" (a near-liquid cement mixture), or should they be caulked with a modern elastomeric sealant? The advantages and disadvantages of each approach should be investigated. Thermal expansion and wind loading as well as environmental deterioration of materials should be considered.
- The need for vertical and horizontal thermal expansion joints may be well understood today, but few early buildings include any such provision. Work should be done to determine if such expansion joints need to be introduced into early masonry buildings and, if so, how this can be accomplished.
- Problems from the condensation of interior moisture must be more thoroughly studied in buildings with curtain wall construction,

some of which have no inner cavity to prevent moisture passage. Continual condensation could cause corrosion to the metal anchorage devices. Condensation may become an increasingly serious problem as buildings are renovated and their use changes (i.e, from office to residential) and as new energy-saving mechanical equipment is added which may not adequately control interior humidity.

• Study of cost-effective and practical ways to apply thermal insulation to masonry walls in early twentieth century buildings must be undertaken. Problems of moisture migration through the walls, increased potential of freeze/thaw problems with masonry, and the aesthetic considerations of adding insulation to interior or exterior wall surfaces should be thoroughly studied.

SELECTED BIBLIOGRAPHY

Architectural Graphic Standards. John Wiley: New York, 1970.

Atkinson, William. An improved skeleton construction, *The American Architect and Building News.* Jan. 9, 1897, pp. 5–6.

Baker, Ira O. *A Treatise on Masonry Construction.* John Wiley: New York, 1889.

Birkmire, William H. *Architectural Iron and Steel.* John Wiley: New York, 1894.

Croly, H.D. The proper use of terra-cotta, *Architectural Record.* Jan. 1906, pp. 73–81.

Duell, John, and Fred Lawson. *Damp Proof Course Detailing.* The Architectural Press: London, 1977.

Eldridge, H.J. *Common Defects in Buildings.* HMSO: London, 1976.

Elements of Stone and Brick Masonry. International Textbook Co.: Scranton, Pa., 1930.

Failure of buildings, *The American Architect and Building News.* April 25, 1896, pp. 36–39.

Freitag, Joseph K. *Architectural Engineering.* John Wiley: New York, 1901.

Fryer, William J. Skelton construction, *Architectural Record.* July, 1892, pp. 228–35.

Hodgson, Fred. *Practical Stonemasonry Self-Taught.* Frederick Drake: Chicago, 1902.

Howe, Malverd A. *Masonry: A Short Text-Book.* John Wiley: New York, 1915.

Jenney, W.L.B. The dangers of tall steel structures, *Cassier.* March 1898, pp. 413–22.

Kenly, W.W. Preservation of materials, *Architectural Record.* Nov. 1903, p. 409.

Kidder, Frank E. *The Architects' and Builders' Pocket-Book.* John Wiley: New York, 1916.

Lynch, Thomas C. *The Masons', Bricklayers' and Plasterers' Guide.* Riggs Printing House: Albany, New York, 1892.

Marsh, Paul. *Air and Rain Penetration of Buildings.* The Construction Press: London, 1977.

Parker, Harry, C.H. Gay, and J.W. MacGuire. *Materials and Methods of Architectural Construction.* John Wiley: New York, 1964.

Pelton, John L. Adjustable facing (marble facing), *The American Architect and Building News.* Nov. 28, 1896, pp. 74–5.

Preservation of steel in tall buildings, *Scientific American.* March 16, 1907, pp. 226–27.

Ramsey, Charles G., and H.R. Sleeper. *Architectural Graphic Standards.* John Wiley: New York, 1936.

Recommended Minimum Requirements for Masonry Wall Construction. Bureau of Standards: Washington, D.C., 1925.

Sturgis, Russell. Stone in American architecture, *Architectural Record*. Oct. 1899, pp. 174–202.

The enemies of structural steel, *Scientific American*. July 6, 1907, pp. 4–5.

The preservation of building stone, *The American Architect and Building News*. March 19, 1887, pp. 142–43.

Webb, Walter L., and W.H. Gibson. *Masonry and Reinforced Concrete*. American School of Correspondence: Chicago, 1909.

Diagnosis and Prognosis
of Structural Integrity

NEAL FITZSIMONS and JAMES COLVILLE

Present deficiencies in historic masonry monuments and buildings may be generally classified as structural, aesthetic, or both. Decisions related to preservation and/or restoration of such structures depend on accurately assessing the existing structural condition and identifying the cause of the deficiency. This paper presents a general methodology for the diagnosis and prognosis of structural integrity in masonry structures. Basic steps in this procedure include preliminary evaluation, on-site investigation, off-site research, preliminary analysis, laboratory and field tests, and structural evaluation. The paper also outlines a deficiency correction process, identifies major areas of needed research, and makes specific recommendations that can be implemented regardless of the structure under study.

Any discussion of the structural integrity of historic structures is complicated by the variety of construction materials and types of construction and architecture encountered. Adding to the complexity is the extreme range of condition and age of historic structures. Associated with existing condition and age are differing concepts of adequacy. Thus, structural problems in the Tower of Pisa or in castle ruins in Scotland, for example, are related to preservation of the existing structure, while in other, more modern historic structures the objective may be restoration to the as-built condition. Finally, a successful (i.e.,

---

Neal FitzSimons *is Principal Engineering Counsel, Kensington, Maryland.* James Colville *is Professor of Civil Engineering, University of Maryland, College Park.*

accurate) diagnosis and prognosis of structural integrity, upon which the identification of proper remedial action hinges, is greatly influenced by limited availability of resources, missing historical information, and limited material sampling.

In view of these complications the main purpose of this paper is to present a general methodology for diagnosis and prognosis of structural integrity in masonry structures. In particular, certain basic procedures and guidelines are presented along with specific recommendations and approaches that can be applied regardless of the particular structure under study. Most of the material presented is drawn from past experiences of the authors and from a recent report to the National Bureau of Standards by the senior author.[1] This report also contains a comprehensive bibliography of pertinent publications.

GENERAL REVIEW OF PROBLEM

That a great majority of historic and other older structures are of masonry construction is, in itself, testimony to the durability of masonry materials, which include burned-clay units, concrete masonry units, and natural stones bonded together with a cementitious material such as mortar. Such construction materials can be manufactured locally and, properly used, provide aesthetically pleasing structures. Rational methods of design of load-bearing masonry structures, based on modern engineering principles, however, have been developed only recently. Before World War II, masonry design was based on a few empirical rules developed from experience gained over hundreds of years of construction dating back to Roman times.

Present deficiencies in historic masonry monuments and buildings may be classified as structural, aesthetic, or both. Structural deficiencies may result from a number of factors. In relatively new construction, problems may generally be attributed to faulty design, including the use of improper or incompatible materials or poor workmanship. In older construction that has survived years of use with limited damage from normal foundation settlement and gradual deterioration, faulty design can be dismissed as the principal reason for recently accelerating decay. Rapidly increasing deterioration of basically sound original designs may be attributed to such factors as excessive loading, accident, human action, and/or significantly increased environmental exposure. Excessive loading includes any loading greater than that considered in the original design, such as sonic booms and vibration resulting from traffic. Accident could include seismic disturbances and other natural phenomena. Human action includes increased tourism, vandalism, and improper remedial treatment. Finally, recent escalations in environmental pollution—acid rain, for instance—also cause significant deterioration.

Assessing the structural condition of an existing building may be a difficult problem, depending on the extent of damage. Even more complex in many cases is identifying the cause of the damage, which is essential to the development of appropriate corrective action. The investigator's basic objective is to arrive at a course of action that will reestablish structural adequacy, and eliminate or minimize the cause of the existing damage without creating new hazards.

Various problems are encountered in the preservation and/or restoration of historic masonry structures. Many of these lie outside the traditional scope of the art and science of engineering. Thus, basic questions relating to the validity of solutions that alter the natural environment, including removal of the structure, protective coatings and enclosures, concealed structural appendages, or partial replacement with simulations, must be addressed. More technical problems—development of structural and architectural restoration criteria, identification of structural analysis methodologies, research to establish appropriate nondestructive testing methods, and research to categorize and identify deficiences—can be resolved with proper resources and support.

Obviously, cooperation and contributions in the development of needed knowledge and philosophy will be needed from a variety of disciplines.

THE DIAGNOSIS

The evaluation of existing structural integrity is only one phase in the overall rehabilitation process, which also includes social, economic, and political considerations. A flow diagram of the evaluation process (Figure 1) should be followed in this process. Major elements of the diagram are described below.

Preliminary Evaluation Plan

The preliminary evaluation plan involves establishing procedures for preparing a project dossier, including methods of organizing information and of documenting and recording data obtained in the diagnostic activities. However, concurrent with organization of the diagnostic effort, pertinent available information concerning the structure should

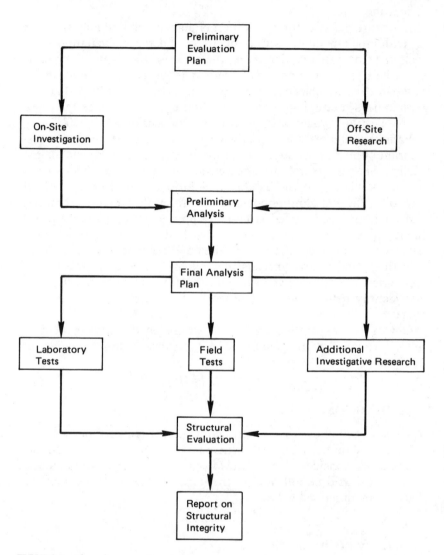

FIGURE 1 Flow diagram of evaluation process.

be collected, including relevant maps, plans and drawings, photos, and other documents.

Site Investigation

A first step in site evaluation is a detailed visual examination of the structure to assess its overall condition and to seek broad clues concerning the severity and possible causes of the damage.

In practice, manifestations of structural deficiencies may be obvious or difficult to detect. Attention should be concentrated on examining rectilinearity of the structure and its apertures. Any rotation or translation of the structure should be investigated. Distress in the major structural components as evidenced by cracking, spalling, or bowing should be noted. Deficiencies providing access to water should be sought, since water penetration can have serious deleterious effects on both structural and nonstructural masonry components. Many of these external deficiencies can be detected by visual examination. More precise observations, if necessary, may be obtained using a variety of measuring and/or surveying instruments.

If possible, each deficiency should be categorized as one of the following types: distress, deterioration, damage, or defect. Distress results from loadings in excess of the structure's design capacity. Deterioration is defined as the result of erosion of structural capacity by environmental attack. Damage is the result of extraordinary loads. A defect is a variation from the intended structural plan that is serious enough to affect structural capacity. In practice it may not be possible to classify the deficiency easily; many deficiencies result from an interaction of these effects.

Randomly located deficiencies complicate the diagnosis. Therefore, from the beginning of the investigation, it is important to identify any patterns of deficiencies. Because cracking is a significant indication of potential structural problems, the location, magnitude, and extent of cracking should be carefully noted and studied.

In masonry construction in general, points of high shear and low moment should be identified and examined, because joints, bearings, and connections are normally the most vulnerable and critical areas of a structure.

Testing Alternatives

Visual site investigation, along with a review of available documents, is intended to permit identification of deficient portions of the structure. Further investigation is necessary if it is prudent to validate observations and assumed or calculated data.

In normal situations where the condition of the construction materials is suspect, it is possible to remove samples from several areas of interest and perform laboratory tests of the properties of the materials. In historic buildings this may not be possible, and in situ, nondestructive measurements and testing will be required to diagnose structural damage. For example, standard surveying techniques can be used in the investigation of most foundation problems. Differential levels and distances accurate to one minute of angle and 0.01 ft (3.048 mm), respectively, are precise enough for most situations. Special equipment can be used to measure vibrations in the structure if they are considered a potential source of distress. Crack sizes can be measured using the simple monocular reticule and feeler gauges. Judgment is required in determining the type and extent of such measurements and the number, size, and location of material samples for further study.

If the cause of the damage is considered to be water penetration and permeance, it may be necessary to use special nuclear instruments to discover the source and extent of the unwanted water and moisture. Testing alternatives may also include nondestructive, full-scale testing of portions of the structure. Such tests can be short-term or long-term. Short-term loading tests are generally more feasible and may consist of static or dynamic load tests. Dynamic tests may be further subdivided into impact and cyclic tests. With the exception of long-term tests involving implanted instrumentation, the scope of full-scale structural tests is usually very limited.

Regardless of the testing to be performed, specific objectives of the tests must be determined well in advance. In general, test objectives may be classified according to the type and condition of the structure—new, extant, distressed, deteriorated, damaged, defective, or repaired.

A research project is under way to assess the applicability of four nondestructive test (NDT) methods currently used for evalution of soil, rock, and concrete to the evaluation of masonry structures.[2] The four NDT methods are: hardness, mechanical pulse velocity, ultrasonic pulse velocity, and dynamic response vibration techniques.

Analytical Techniques

The strength, stiffness, and stability of a structure must be determined from the data gathered through investigation and testing. This information, at best, will be incomplete. Analysis of a structure is based on known geometry and material properties and characteristics, from which allowable stresses are estimated. The analysis produces values of allowable loads or factors of safety with respect to design loads. Often, funds, time, or circumstances lead to analyses based on incomplete data. In such cases certain statistical techniques may prove helpful when combined with real but incomplete data.

The basic components in masonry construction will be the masonry units, mortar, ties, and reinforcements. Masonry units vary widely in strength and durability. This is also true of mortars, which, until fairly recently, were lime-based with no portland cement. The compressive strengths of mortars can be extremely variable, ranging from less than 100 psi to more than 3,000 psi. Brick units, depending on their new materials and manufacturing processes, can have ranges in compressive strength from 1,500 psi to more than 16,000 psi. Structural tiles have similar variations, with compressive strength varying from 1,100 psi to more than 10,000 psi. Concrete block units, although highly controlled in recent years, will also display significant ranges in properties, as will stone units.

Some structures may have been designed using the rule-of-thumb approach, and an analysis using modern techniques may indicate considerable latent capacity. Including nonstructural elements in the analysis may also reveal significant influences on structural stability.

Structural Evaluation

Final structural evaluation should include a description of existing deficiencies along with a discussion of the causes of such deficiencies. The present condition of the structure should be defined, with descriptions of both sound and deficient sections. Finally, the conclusion should be presented along with substantiating documentation concerning the as-is capacity.

THE PROGNOSIS

After the diagnosis of structural integrity, an equally difficult but necessary part of the overall process is to determine recommended remedial action. The necessity of such action and the type of action recommended depend on accurate prognosis of structural adequacy with and without corrective action. These prognoses will depend on the combined effects of reducing vulnerability and improving the structure. The remedial plan, which includes the definition of tasks and time and cost estimates, should have these two objectives clearly before it. A satisfactory solution may involve either or both of them.

A typical deficiency-correction process is illustrated in Figure 2. Elimination of the cause of the problem may be a sufficient solution. Elimination of the effect of the problem may not be a lasting solution. Minimizing vulnerability to the cause is a valid objective. But finally, minimizing or eliminating both cause and effect provides the optimum solution.

Care is needed if part of the solution is to create a new environment

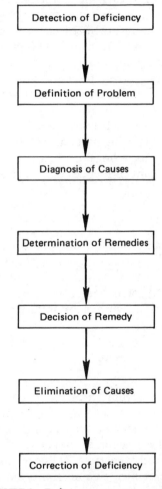

FIGURE 2 Deficiency-correction process.

for the structure. Changes in ambient internal moisture and temperature may create more problems than are solved. In particular, new conditions resulting from improving the structural capacity may in themselves lead to secondary problems that hopefully will be minor. All changes should be carefully evaluated with respect to their effect on the structure's foundations. Addition of members, walls, materials, etc., to improve the capacity of the superstructure may cause redistributions of loading with subsequent detrimental effects on the substructure. All major modes of foundation failure, subsidence, rotation, and translation should be considered in the final report.

Finally, after completion of remedial work, it is important to monitor the performance of the structure periodically to assess the impact of the intervention on its physical life.

REFERENCES

1. FitzSimons, N. Structural Evaluation Guide for Building Rehabilitation. Report submitted to National Bureau of Standards, July 18, 1979.

2. Noland, J.L., and Atkinson, R.H. An Evaluation of Non-destructive Test Methods Applied to Masonry. Proceedings, Conference on Research in Progress in Masonry Construction, March, 1980, Marina Del Ray, California.

Photogrammetric Measurement and Monitoring of Historic and Prehistoric Structures

THOMAS R. LYONS and JAMES I. EBERT

Photogrammetry is measurement using photographs or other images of re-
motely sensed electromagnetic data. It provides architects, archeologists, and
others concerned with prehistoric and historic cultural resources with cost-
effective means of recording and monitoring structures, monuments, and other
cultural sites. This presentation sets forth the basic principles of photogrammetry, concentrating on the use of vertical aerial photographs and terrestrial
recording of structures with a ground-based camera. Controlled stereo photographs can be used as the basis of a number of photogrammetric products,
including photogrammetric maps, orthophotos, and combinations thereof. Repetitive aerial and terrestrial photographs of a site, and products derived from
them, can be compared to allow the monitoring and documentation of inevitable change in cultural resources. Methods by which this is accomplished
are illustrated by reference to a number of ongoing monitoring experiments.
Considerations dictating frequency of monitoring and precision of measurement are discussed, as is the multiuse nature of photogrammetric products.
A wide-spectrum approach to monitoring cultural resources that integrates
the use of past data sources such as historic photographs and maps, present-
day photogrammetric products and data, and possible future techniques such
as microwave scanning and holography, is necessary to ensure the conservation
and integrity of our historic cultural resources.

Thomas R. Lyons *is Chief, Remote Sensing Division, Southwest Cultural Resources
Center, National Park Service, Albuquerque, New Mexico;* James I. Ebert *is Archeologist, Southwest Cultural Resources Center.*

242

This paper is concerned with the methods of aerial and terrestrial photogrammetry and their application to the measurement and monitoring of historic and prehistoric architectural structures. The purpose of these procedures for documenting the condition of building fabric is to provide a basis for quantitative measures of alteration or deterioration over time and thus to provide data for informed decisions on maintenance, preservation, or restoration.

The photographic products of calibrated aerial and terrestrial cameras are historic documents of value to historical architects and/or archeologists. Under certain circumstances of acquisition they provide a data base that can be used for comparative purposes, for faithful restoration or reproductions.[1]

Repetitive controlled imagery has the added function of establishing a measure of the rate of change in building fabric and consequently a foundation for predicting future deterioration. It therefore allows the foreknowledge necessary to prevent or mitigate harmful effects.

INTEGRITY OF HISTORIC STRUCTURES

The value as heritage and the scientific significance of historic and prehistoric architectural structures depend to a great extent on the integrity of the fabric of which they are composed. The term "integrity" as used here does not imply the absence of change. Change in materials over time is inevitable, and humanly induced changes in structures occur regularly and can add interest and value to historic properties. A historic structure has integrity when the process of change through time can be documented and when the nature of the changes and their causes have been determined. Photogrammetry, a remote sensing technique involving accurate measurement from photographs or other images of electromagnetic data, is a powerful tool for the historian, historical architect, or archeologist wishing to document the history and causes of change in architectural resources.

BASIC PRINCIPLES OF PHOTOGRAMMETRY

Photogrammetry

Photogrammetry, in its strictest sense, is measurement using photographs. A single aerial photograph for instance, is a representation of a three-dimensional scene reduced to a two-dimensional format. While it faithfully presents scaled measurements for each plane in the scene parallel to the film plane, it also contains a number of distortions. One of these is occasioned by the fact that the relationship between the actual size of an object photographed and its size on the film depends on its distance from the camera (focal length of the lens being held equal). Even in a vertical aerial photograph, this causes differences in scale between an object on the top of a mountain and an object at the bottom of a valley since they are different distances from the camera.[2] Radial or relief displacement is another effect of forcing a three-dimensional scene onto a planar film. For example, if a tall object such as a masonry obelisk appears at the exact center of a vertical aerial photograph, only its top can be seen; but if it appears anywhere else in the image, not only its top but also one of its sides may be seen (Figure 1). A single, monoscopic photograph containing such distortions obviously is not an accurate representation of reality.

Cameras used in photogrammetry are classified by the platform on which they are placed. Space or airborne cameras are aerial cameras, while those resting on the earth and pointed horizontally rather than vertically are terrestrial cameras. The difference in the orientation of these two types of cameras allows the architect and archeologist to derive accurate representations of structural elevations and plans from both a vertical and horizontal perspective.

Stereophotography

The use of stereophotography turns the distortions inherent in monoscopic photos into advantageous information.[2] Two sorts of measurements that can be made from a stereo pair of aerial photographs are parallax measurement and radial line plotting.

Parallax measurement makes use of the radial displacement inherent in each of the photographs in a stereo pair. The difference in distance between the base of an object and the top of the same object in each pair is measured; if the scale of the photographs is known, the height of the object measured can be determined. The principle allows the measurement of the relative heights (z coordinates) of each point in a photograph and is the basis of topographic mapping.

The accuracy of the x, y, and z dimensions derived through parallax measurement and radial line plotting from a stereo pair depends on the quality of the photographs themselves. The film plane must be extremely flat to minimize uncontrolled scale variations within the image, and the lens axis must be accurately perpendicular to and centered over the film plane. As a consequence, photographs taken with standard cameras are not generally suitable for mapping, and metric cameras are used instead. Unlike "snapshot" or even higher-priced SLR

1:1920   3-004

FIGURE 1 Relief displacement illustrated by an aerial photograph showing an obelisk at Chalmette National Historical Park in Louisiana (upper left corner). If this obelisk had appeared at the exact center of the photograph, only its top would have been visible. Since it was at the edge of the photograph instead, the sides are clearly visible.

cameras, metric cameras are designed to take extremely accurate photographs. They are collimated to insure that the axis of their lens is perfectly perpendicular to the film plane, their platens bear fiducial marks so the exact center of the lens axis can be determined on the photograph, and they are precisely calibrated in a laboratory prior to sale. While metric cameras range in price from $10,000 to $60,000 or more, a number of firms maintain and use them regularly and will enter into contracts for terrestrial and aerial photogrammetry.

True horizontal mapping of structures or features from vertical aerial photographs is accomplished through radial line plotting. Radial line plotting makes use of the geometric fact that any point in space can be defined by sighting it from two known reference points or stations and recording the angles to the unknown point. In mapping, this procedure is referred to as triangulation. Radial line plotting using a stereo pair of photographs is usually done with a radial plotter or photogrammetric plotter. The instrument uses the center of each vertical photo as a known station and allows the interpreter to position a line from this point through each point of interest on the photograph. Such points might trace the outline of a structure or site feature. A mechanical linkage provides a means of moving a tracing pencil, which creates, in effect, a matrix of points of intersection that planimetrically represents the location of features on the ground.[3] Although such data can be obtained through a field survey using a transit or alidade, radial line plotting from aerial photos is many times faster and more economical.

Accuracy, Precision, and Scale

Accuracy is the closeness of a measurement to the real world; it depends on the scale at which photographs are taken and interpreted and the scale at which these measurements are plotted. In general, the larger the scale of a photo negative, the more accurate a final photogrammetric product will be. Precision refers to the distribution of the values of a number of samples of a single parameter about the mean value—in other words, the replicability of a measurement. Precision in photogrammetry depends primarily on distortions or errors associated with the photography, the plotting device used, and the plotter operator. These errors in turn depend to a certain extent on the scale of imagery and map plotting, especially the variation from one interpretation of a stereo pair to the next by the plotter operator. Accuracy is important in photogrammetric mapping; accuracy and precision are vitally important in the sorts of comparisons that constitute photogrammetric monitoring of architectural targets.

The scale at which controlled photographs of a site or structure are taken must be decided on the basis of a balance between the constraints imposed by reality (economics and the capabilities of equipment) and the needs of the researcher (problem orientation). It is not practical, nor is it very definitive, to ask for "as large a scale as possible," and the scientist or manager who does so probably has not thought seriously about his need.

If photogrammetric products and data are to be used for accurate reconstruction of a structure, the accuracy to which building materials can be fabricated will be a criterion for determining scale. When a structure is being recorded for monitoring purposes, questions of how much its fabric changes cyclically (diurnally, for instance) and what sort of change would signal impending collapse or irreparable damage must be asked. Answering these questions may require relatively high accuracy. Theoretical or scientific questions, on the other hand, may not be so demanding. If one were to survey the ethnoarcheological literature for predictive formulae for determining the population of structures, for instance, one would find that the variance in estimates and observations is high enough to make the use of any photogrammetry at all questionable.

Photogrammetry need be no more accurate than the problems to which it is to be applied warrant. In applying this rule to the determination of scales, it should be remembered that to double the accuracy at which a map can be plotted, one must double the photographic scale and therefore quadruple the area that must be photographed and interpreted. Necessarily, there is a corresponding increase in cost.

Precision or replicability of photogrammetric results, which is necessary for comparing periodic data from a specific site or structure, is a problem. By permanently marking and reusing control points and employing the same metric cameras, camera positions (difficult from the air), and plotting equipment, a certain amount of comparability between maps can be ensured.

Another problematic element is inconsistency on the part of the plotter operator. Photogrammetric maps are interpretations of photographs, accurate ones to be sure, but including different amounts of detail depending on the inclination of the human interpreter. This is especially apparent with contour lines drawn on two maps of the same target. Topographic contour lines are very difficult for the same operator, using the same imagery, to duplicate exactly. Point locations, on the other hand, can be duplicated by different interpreters with surprising accuracy. For this reason, it is advantageous to locate specific points to be used as the basis of monitoring a structure rather than attempt to use a contour map of, say, the face of a wall as monitoring data. This requires the selection of conspicuous, relocatable points (along the top of a wall, at the corners of windows, etc.) and may require that permanent markers be placed at these points for some structures.

Orthophotography

An orthophoto is essentially a photographic planimetric map. It is produced by a machine that not only does radial line plotting, but also uses parallax measurement as the basis of changing the scale of very small segments of a monoscopic photograph as these small segments are exposed on a film. This corrects for scale differences throughout the photograph and creates a planimetrically correct print of a scene. Orthophotos are more useful than planimetric maps for many purposes because there has been no selection by an interpreter of what points in the scene are of interest; all points (within the limits of resolution of the photograph) have been shown, and details not available on a map can be distinguished at a later date.

A combination of an orthophoto and a photogrammetric line map can be made by first compiling an orthophoto and then using the same controlled stereo pair(s) on a photogrammetric plotter to produce a topographic contour map at a desirable scale. The line map is then photographically overlain on the orthophoto (in negative form—white map lines against the darker orthophoto, Figure 2). Such a product combines the advantages of both of its parts: Landmarks and other details can be found and used for measurement on the orthophoto, while quantitative, three-dimensional data are available from the topographic map lines. An orthophoto/topographic map combination is especially useful as an aid to planning survey, excavation, or conservation activities in the field.

Ground Control

To relate the three-dimensional coordinate values of each point photogrammetrically measured in a stereo model to other measurements from the real world, it is necessary to mark and measure a few control points on the ground or target to be photographed before the photos are taken. Ideally, three or more horizontal and four or more vertical control points are set and marked so that they will be visible in both of the photos forming a stereo pair (Figure 3). The distance between the horizontal control points is measured, and the difference in elevation of the vertical control points is established. This allows the operator of the photogrammetric plotter to determine the precise scale of the photos and insert them (printed on glass plates rather than paper, and called diapositives) in the plotting machine. If control is not set prior to taking stereo photographs, certain conspicuous points on the ground can often be "photoidentified" in the image and then located and measured on the ground, serving as control points after the fact (Figure 4).

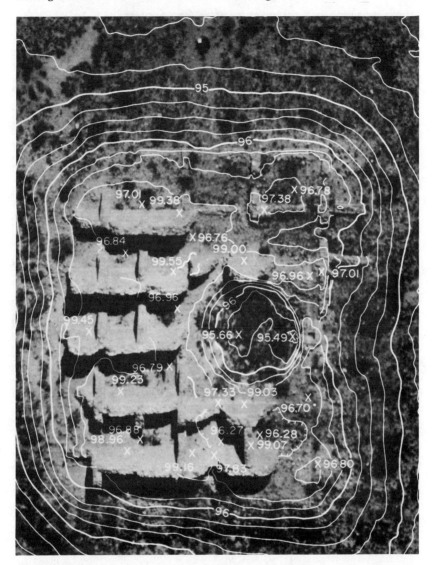

FIGURE 2 An orthophotograph of a Pueblo-III Period masonry site in San Juan County, New Mexico, with superimposed topographic contours. The orthophotograph is essentially an aerial photo corrected to show a planimetric view of the site while also conveying details inherent in a photograph. Topographic contours were compiled using a stereo plotter and photographically superimposed over the orthophoto, thus supplying metric information as well. Such renditions are especially useful as base maps for planning and executing fieldwork. SOURCE: Koogle and Pouls Engineering, Inc., Albuquerque, New Mexico.

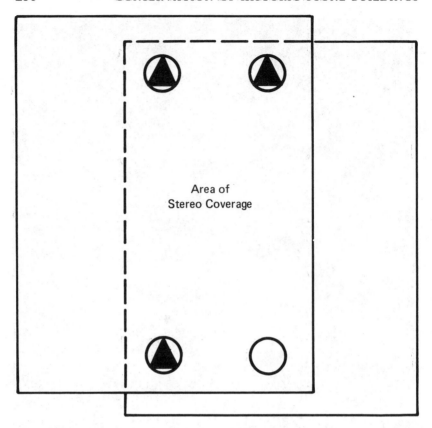

FIGURE 3   Minimal horizontal (triangles) and vertical (circles) control points required to be marked and measured in a stereo model. When flight lines are being used as the basis of mapping, it is sometimes possible to use fewer control points and still maintain photogrammetric control by "bridging." Control points are located in the field prior to flying and are marked with stakes and plastic panels so that they are visible in the photographs taken later. The distances between horizontal control points, and the differences in elevation between vertical control points, are then measured. Control points can also be tied to known data points if desired.

Aerial (Far-Range) vs. Terrestrial (Close-Range) Photogrammetry

Aerial photographs are the staple for far-range photogrammetry. They are used every day by engineers, planners, geologists, cartographers, and many others for the development of topographic and planimetric data. Aerial photographs are taken in scales ranging from about 1:200,000 (small-scale photography) to 1:500 (large-scale photography). Equipped

FIGURE 4 An aerial photograph of the Barbourville Mansion in Virginia, a brick structure designed by Thomas Jefferson and destroyed by fire in 1844. The L-shaped white markers in the corners of this scene are ground control panels, set prior to overflight. If these panels had not been set, it would have been necessary to photoidentify points and measure the distances between them later. Examples of points that are well defined and could be used for "subsequent control" are shown at A and B.

with the most commonly used aerial cameras, airplanes cannot fly slowly and low enough to take larger images. Using a first-order plotter, a photogrammetric map scale of 1:100 (1 in. = 8.33 ft) can be obtained using 1:1,000 scale imagery at a contour interval of as little as 0.1 ft (3 cm). The use of aerial photography for photogrammetry offers the possibility of mapping very large areas inexpensively but at low resolution and smaller areas more precisely up to this limit. Because of the design of most aerial plotters, oblique aerial photography is of little utility in photogrammetry.

252

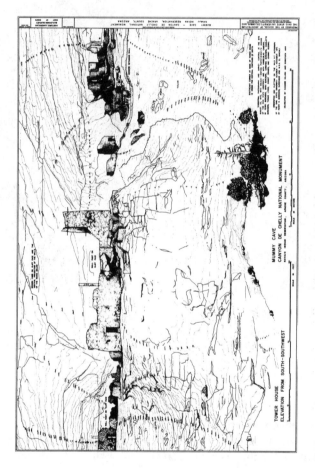

FIGURE 5 Photogrammetric elevation, with added detail, of Mummy Cave in Canyon de Chelly National Monument, Arizona. This elevation was compiled using terrestrial photogrammetric photographs, which are collected by a camera with a horizontal rather than a vertical lens axis. SOURCE: Perry Borchers, The School of Architecture, Ohio State University, Columbus.

Terrestrial photogrammetry uses photographs taken with a terrestrial photogrammetric camera or a phototheodolite, the ground-based equivalents of an aerial camera. For terrestrial photographs, the film plane of the camera is arranged to be parallel with the structure being photographed. Control is as important in terrestrial photogrammetry as it is in measurement from stereo vertical photos. Terrestrial photogrammetry results in maps or drawings showing planimetric detail in the x and z (or y and z) planes, with topographic point or contour data in the y or x plane (Figure 5).

Most terrestrial photogrammetry in the United States is done with photogrammetric plotters designed primarily for use with vertical aerial photos. There is little latitude for image tilt or scale differential with such machines, making camera setup very critical. Some European plotters are designed for use with terrestrial stereo photographs, however, and allow the plotting of planimetric detail from oblique stereo photos. These instruments have been coming into use in the United States in recent years. They can transform terrestrial photographs into a "vertical," and since they obtain imagery at much larger scales than possible from an aircraft platform, they can be used to advantage in large-scale drawings of historic and prehistoric architectural features.

Digitizing Photogrammetric Data

The products and representations of photogrammetric data discussed above have all been graphic. Another useful way to represent photogrammetric data is by digitizing three-dimensional point coordinates. Many modern plotting devices incorporate a digitizer that records coordinates chosen by the plotter operator on magnetic tape or computer cards for later analysis and manipulation. Points may be chosen at regular intervals along the tops of walls or the outer perimeter of a building (from vertical, etc.), or at intervals on the surface of a wall or structural member (with terrestrial photographs). The operator may also become an interpreter, when properly trained, and identify only those areas of significant change in, say, the elevation of the top of a wall, that should be measured. Figure 6 is an aerial photogrammetric map of a prehistoric pueblo structure at Chaco Canyon National Monument with x's marking some of the points digitized during photogrammetric plotting.

Digital photogrammetric data can be used in a number of ways in structural documentation. The simplest of these is graphic representation of a structure as a matrix of points connected by lines. Such a

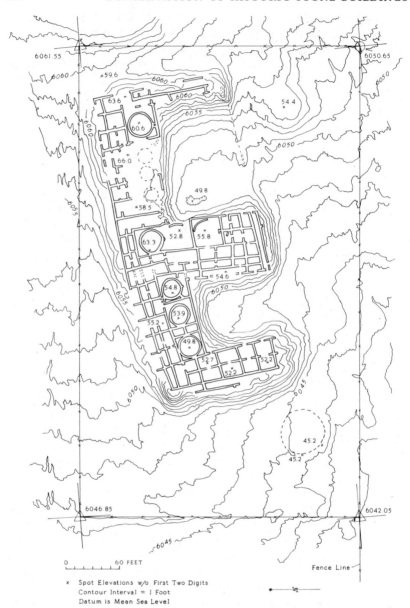

FIGURE 6 Photogrammetric map of Kin Bineola Pueblo, Chaco Canyon National Monument, New Mexico. The elevation figures on room floors and along the irregular tops of walls (numbers not shown) were digitally recorded for computer manipulation. SOURCE: Drafted from photogrammetric data by Jerry Livingstone, Southwest Cultural Resources Center, National Park Service.

representation can be rotated automatically by a computer to create views from any direction or angle or elevation (Figures 7 and 8). Digitized data from a drawing of one map or elevation can be compared with subsequent data as the basis of monitoring a structure.

Engineering and Art

Just as different scales, and hence accuracy and precision, are acceptable or required for different needs and problems, the form of the final product produced through photogrammetry may vary according to the applications to which it is to be put. One of the most often noticed differences is the level of "artistic refinement" of terrestrial photogrammetric elevations of structures. The first product to emerge from a photogrammetric plotter is a manuscript map or drawing. It shows all planimetric detail and three-dimensional coordinates or contours chosen for interpretation, but often is drawn by a pencil or single-width plotter pen. The lines on some manuscript maps are thus often

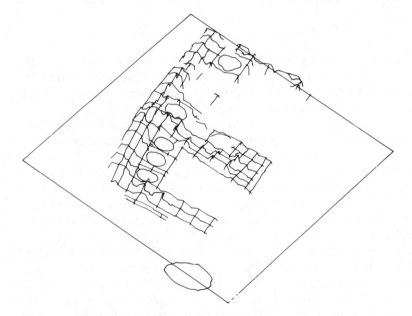

FIGURE 7 A perspective line drawing of Kin Beineola Ruin produced by a computer from digitized wall height data. It should be noted that "hidden lines" have been shown in this version, conveying more information than if they had been hidden, but also tending to confuse casual viewers. Computer programs exist by which such hidden lines can be removed.

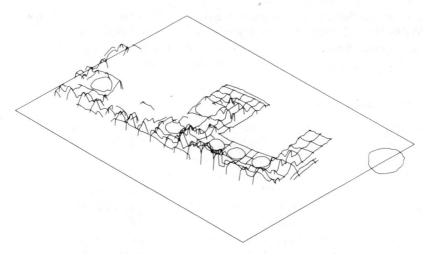

FIGURE 8   A computer perspective line drawing of Kin Bineola Ruin from a different angle. When combined with room-floor data that indicate the depth of rubble fill, such perspective information can serve as the basis of computer reconstruction of a structure (see page 269).

thin, difficult to reproduce, and far from aesthetically pleasing (Figure 9). Photogrammetric firms usually compile a second product from the manuscript map by drafting or scribing the lines represented on the manuscript map for greater clarity or reproduction.

Drafted maps contain the same information as manuscript maps and are far more useful for publication. Interestingly, during some early experiments in mapping and monitoring structures at the Remote Sensing Division of the Southwest Cultural Resources Center it became apparent that even drafted maps and drawings were not acceptable, or at least not very exciting, to architects. For their purposes, a third-stage product is necessary, and for want of a better term we shall call it an "artistic rendition." Such a drawing contains not only photogrammetric data and cartographic information, but information about how a structure actually looks. Unlike a photogrammetric map overlain on an orthophoto, such information is often idealized, conventionalized, and selectively added. Two elevation drawings of the same structure (Figures 10 and 11) serve to illustrate the differences between such products.

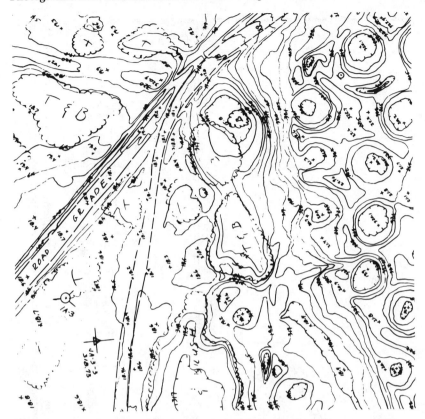

FIGURE 9 Manuscript map of a portion of a historic, abandoned Hidatsa Village at Knife River Indian Villages National Historic Site in North Dakota. The round, depressed areas are the outlines of excavations over which earth lodges were built by the Indians. Note the unsharp and difficult-to-reproduce lines, typical of the first stage of photogrammetric plotting done with a pencil or similar instrument fitted into a servo-mechanism connected with a photogrammetric plotter. SOURCE: Koogle and Pouls Engineering, Inc., Albuquerque, New Mexico.

THE ECONOMICS OF ARCHITECTURAL PHOTOGRAMMETRY

Cost Effectiveness

The size and detail of the target of aerial photogrammetric mapping are the prime factors in determining cost effectiveness. Aircraft setup time, ferry time, time over target, setting of ground control, film and film processing, and laboratory and plotter time are all cost elements

258

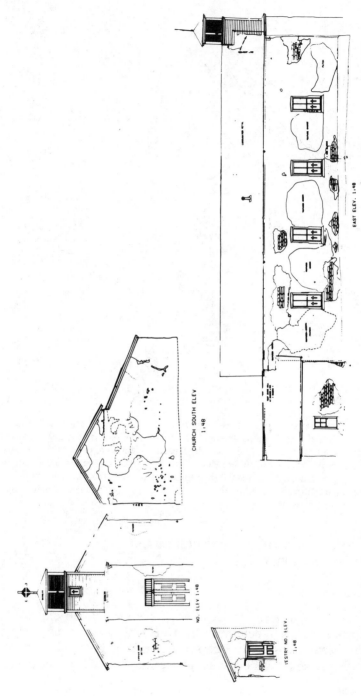

FIGURE 10 A drafted terrestrial photogrammetric drawing taken from a plotter manuscript of the San Juan Church, Lincoln County, New Mexico. This elevation was derived directly from a photogrammetric plotter, and delineates only measurable information. SOURCE: Koogle and Pouls Engineering, Inc., Albuquerque, New Mexico.

259

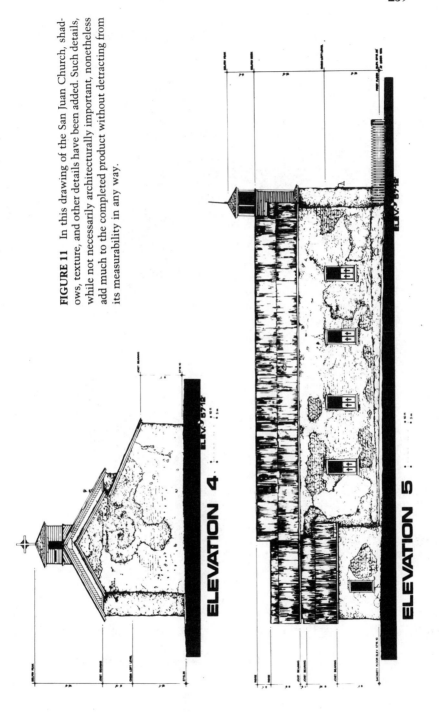

FIGURE 11 In this drawing of the San Juan Church, shadows, texture, and other details have been added. Such details, while not necessarily architecturally important, nonetheless add much to the completed product without detracting from its measurability in any way.

260

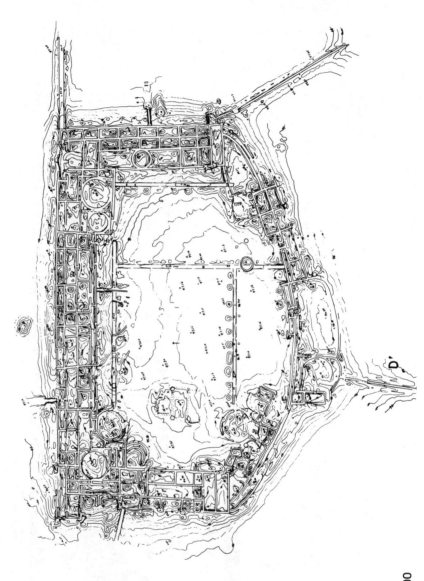

SCALE: 1:1000

FIGURE 12 An aerial view and photogrammetric map of Pueblo Alto, a Pueblo-III Period ruin at Chaco Canyon National Monument in New Mexico. The photographic imagery was produced after the first season of excavation. SOURCE: **Koogle and Pouls Engineering, Inc.,** Albuquerque, New Mexico.

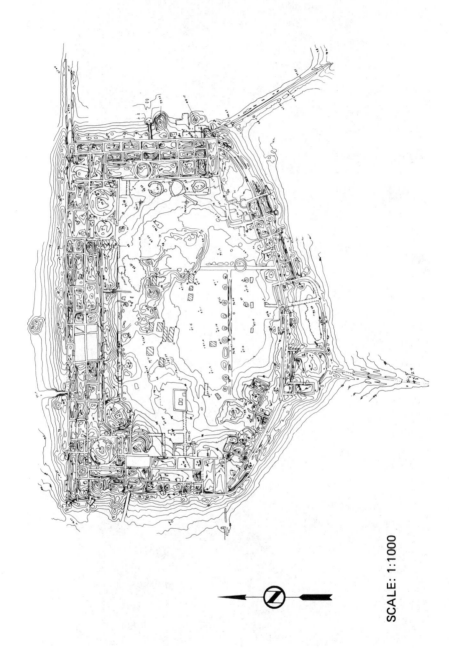

SCALE: 1:1000

FIGURE 13 An aerial view and photogrammetric map of Pueblo Alto following the second summer season of excavation there. New topographic and planimetric data illustrate the progress of the excavation and the change in the structure and its surroundings. These new data were added to the original map [Figure 12, lower portion] by superimposition and did not require a total remapping effort. Archeologists are one of the most destructive forces affecting archeological sites, and such phased mapping during excavation is vitally important. SOURCE: Koogle and Pouls Engineering, Inc., Albuquerque, New Mexico.

to be considered. In general the larger the target or the number of targets covered during a single mission, the more cost effective the mapping method.

It is much more difficult to determine the cost effectiveness of close-range photogrammetry. Again, however, the more the required detail, the more apt the method is to provide savings over manual or standard surveying and recording techniques.

APPLICATIONS RESEARCH

Monitoring

Photogrammetry, as illustrated in this paper, has proved an efficient means of accurately recording architectural data and is being used experimentally for monitoring historic and prehistoric sites and structures. We would like to suggest, however, that photogrammetry in its strictest sense—the use of controlled photography taken with metric cameras and converted maps and elevation drawings or other photogrammetric products—will be supplemented in the future by a number of other remote-sensing measurement methods. Because structural monitoring is a historic process, entailing the use of previously collected baseline data and comparison with subsequent data sets, it is vitally important that future means of recording and monitoring be compatible with past photogrammetric data bases. The first step in this process was initiated at Pueblo Alto in Chaco Canyon National Monument, a ruin of the Pueblo-III period occupied during the eleventh and twelfth centuries A.D.

After the masonry walls in Pueblo Alto were stripped of their overburden of sand and building debris, controlled aerial photographs were taken and a detailed map of the exposed structure was plotted Figure 12). Following excavation and the exposure of additional architectural features and after reidentifying and panelling of the original ground control, the site was again overflown. This imagery was placed into the plotter and projected onto the original map. It was not necessary to remap totally the target but simply to add new details or modify old ones (Figure 13).

This additive process suggested the feasibility of projecting past baseline data onto the original photogrammetric map or drawing and checking for any changes or modifications to the original draft. Experiments have begun to determine the type, degree, and frequency of change due to natural or human impact on fragile masonry structures. Plans and elevations of historical Barbours Mansion, designed by Thomas Jefferson and burned in the mid-1800s, and of the prehistoric Tower

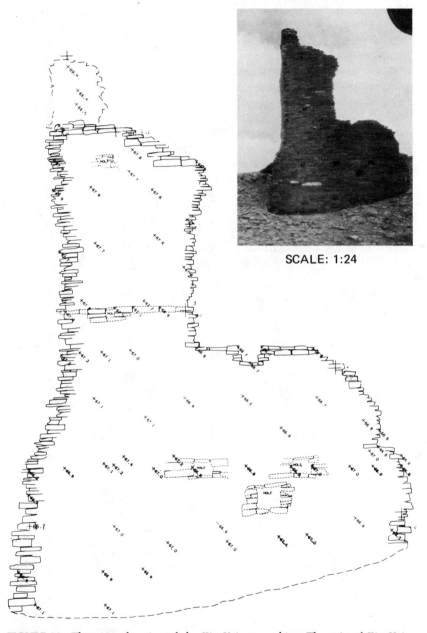

SCALE: 1:24

FIGURE 14 The west elevation of the Kin Ya'a tower kiva. The ruin of Kin Ya'a, originally built of native sandstone, lies within Chaco Canyon National Monument in New Mexico and probably dates to A.D. 1000–1175. Variations in the wall surfaces are measured against a vertical datum plane, and edge details are carefully drawn in by the plotter operator. SOURCE: Koogle and Pouls Engineering, Inc., Albuquerque, New Mexico.

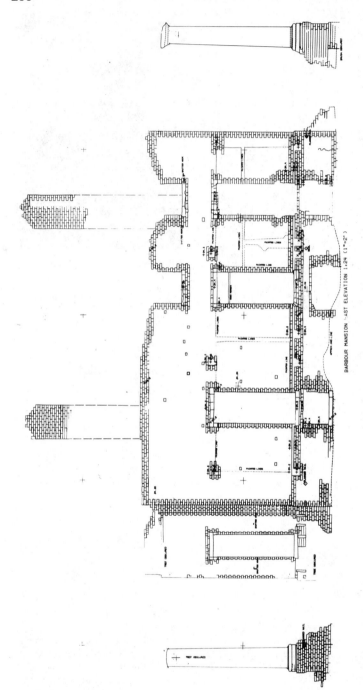

FIGURE 15 East elevation of Barbourville Mansion in Virginia, produced with controlled stereoterrestrial photogrammetric imagery. Only edge details, such as the outlines of separate bricks around windows, have been included here for clarity. SOURCE: Koogle and Pouls Engineering, Inc., Albuquerque, New Mexico.

Kiva at the Kin Ya'a site in New Mexico have been produced and will serve as the baseline data for future monitoring (Figures 14 and 15). Both controlled line drawings and digitized point data derived by photogrammetric means can be used in this method of monitoring structure change.

The frequency of monitoring these features by photogrammetric means is dependent on rate of change. To determine this, aerial and terrestrial photography will be reacquired (for example, every 6 to 12 months). Further, it is anticipated that planned mining activities in the Kin Ya'a area will have a negative impact, and during this activity the periodic rechecks will be more frequent. It is obvious, of course, that effects of earthquakes, fires, or other disasters will be checked as soon as possible.

One monitoring method that is practicable today is the use of electronic distance measurement (EDM) equipment for the repetitive measurement of points on a structure. EDM devices make use of either a laser beam or microwave emitted from a transmitter, reflected from a prism or a point on the object to be measured, and then received at the transmission point. The time taken for the beam to reflect and return is accurately measured and, when atmospheric pressure, temperature, and humidity are corrected for, reveals the exact distance from the EDM to the target. Such equipment is used widely for surveying today, and typical accuracies are ±0.5 cm in 1 km.

Such devices could be used to measure the distance from a fixed station to a number of fixed points on a structure. Points to be measured would be located with respect to their importance in the integrity of the structure—that is, points where a dimensional change would foretell significant alteration or collapse. Such points could be unobtrusively provided with sockets into which a standardized prism would be fitted. A measurement would then be taken from the permanent instrument station (perhaps a socket into which the instrument would always be fitted for measurement) and the prism moved manually to the next point. While more field time would be required for such measurement than with aerial or terrestrial photogrammetry, far less laboratory time (photo developing, instrument setup and plotting, etc.) would be expended.

Many monitoring situations may not require the measurement of the potentially large number of three-dimensional point coordinates available in stereo photos, and EDM monitoring of a relatively small number of highly critical points on a structure or site could greatly speed the monitoring process. It would be feasible, for instance, to monitor a small number of points on a structure each day, or conceivably even several times a day, if this were deemed necessary. In addition, data collected with EDMs would be compatible with point data plotted photogrammetrically.

Microwave Scanning

A microwave scanner, such as side-looking airborne radar or synthetic aperture radar, is essentially an EDM that scans the entire surface of a target area systematically and compiles a matrix of data on the distance from the scanner to the target. Although one commonly sees this matrix represented as a pseudophotographic product, it is recordable in digital form as well. If properly controlled, the digital output produced by a ground-based microwave scanner could have certain advantages over camera photogrammetry. A computer could produce almost instantly a comparison map of two digital point matrices recorded from the same site or structure at different times; it is possible that pattern recognition programs could be developed that would register the two data matrices automatically with manual recourse to control points.

Ground-based microwave scanners that would allow the scanning of specific structures with sufficient accuracy and resolution have not been developed, and a prototype program probably would cost millions of dollars. But this does not negate the possible use of microwave scanning to monitor structures and sites. If we are serious about developing more efficient and effective methods for dealing with our historic heritage, a preliminary feasibility study might well be in order.

Holography

An intriguing process using laser technology for the measurement of physical objects, and the three-dimensional reproduction of their form, has appeared recently in the guise of holography. This technique entails focusing a split laser beam at an object from two directions; reflected light is then exposed on a photographic plate. Projecting a laser beam or other coherent light back through this plate, which contains no obvious image when viewed in normal light, reconstructs a "wave front" image of the object in three dimensions. The hologram can be viewed from different directions, and parts of the object that are hidden from one angle appear from others.

Holography has been used to record historic artifacts at the Smithsonian Institution and might offer a means of recording and monitoring historic structures as well. Before this could be accomplished, however, some startling jumps in laser technology would have to occur. To the present, holograms have been made only of relatively small objects, the largest being humans. The only lasers with enough power to record larger things are those currently being experimented with by the military for purposes far removed from historic preservation. Another problem with holography for the measurement of large structures is that during the exposure of the hologram any movement within the scene (even on the order of the size of a wavelength of light) will disturb or ruin the plate. Ever-changing historic structures might be difficult to stabilize this perfectly for even an instant. Nonetheless, we predict that holography will become a useful aid to recording and monitoring structures and cultural resources in general.

Reconstruction

Digitized three-dimensional photogrammetric data can be used for more than simple representation of a structure or comparison with baseline data for monitoring purposes. In many cases the archeologist or cultural resource manager is concerned with structures that already have suffered considerable destruction or fabric change. Many sorts of theoretical statements are made on the basis of the forms of structures. These include population estimates, assumptions of the importance of communities in a trade network, and guesses about the degree of affluence or "refinement" of occupants. When making such assumptions it is important that one know the original form of the structure, which is difficult to divine in certain instances. At Chaco Canyon National Monument in northwestern New Mexico, for instance, a matter of conjecture is the number of stories that pueblo ruins originally had. An experiment under way at the Remote Sensing Division may help to solve this problem.

Pueblo Alto (Figures 12 and 13), will serve to illustrate this approach. As previously mentioned, the aerial imagery obtained served as the basis of photogrammetric mapping. During the course of this work, some 7,000 points were digitized in three dimensions and recorded on computer cards. Points along the tops of walls were digitized at each significant change in wall height, and other points were chosen within exposed rooms to establish the depth of rubble fill. Using only the wall-top points, a model of the ruin can be reconstructed by the computer.

These wall-top points are being combined with data on rubble fill and information derived during test excavation in an attempt to reconstruct the configuration of the original structure. While computer reconstruction cannot produce a single "true" picture of the original configuration of a pueblo, it can help suggest plausible alternative

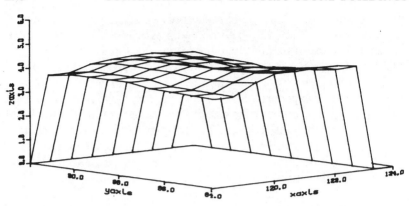

FIGURE 16 Computer perspective drawing of a small portion of the Pueblo Alto ruin. This drawing, constructed with a 10 × 10 mesh data resolution, shows little detail. SOURCE: Joe McCharen, Civil Engineering Research Facility, University of New Mexico.

designs upon which theoretical statements can be based (Figures 16, 17, and 18).

SUMMARY

Both aerial and terrestrial photogrammetric techniques are invaluable for documenting historic and prehistoric masonry or other forms of architecture in plans and elevations with a range of detail and accuracy. As with any procedure, of course, there are limitations to the usefulness

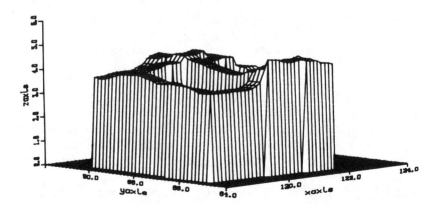

FIGURE 17 A more detailed, 50 × 50 mesh resolution computer perspective drawing of the same portion of Pueblo Alto appearing in Figure 16. Note that "hidden lines" are actually hidden in this rendition. Wall thickness and the irregular profiles of the partially ruined walls are apparent.

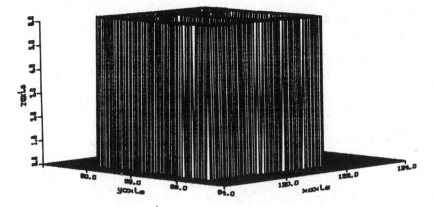

FIGURE 18 Employing digitized wall-top profiles such as those appearing in Figures 16 and 17, as well as room-fill data, the computer has reconstructed the four-room section of Pueblo Alto as it may have appeared when in use. While this experiment is obviously on a small scale, reconstructions of entire prehistoric or historic structures could be postulated using similarly handled digitized photogrammetric information.

of available technological methods for assessing conditions and changes and for establishing baseline or quantifiable data upon which to formulate decisions for preservation efforts.

Digitized photogrammetric point data in conjunction with computer programming gives the architect or archeologist a three-dimensional tool for analysis of structures. Digitized structural data also are useful for monitoring change or the effects of human or natural forces on fabric.

Research in the application of microwave scanning, holography, and three-dimensional, computer-aided structural reconstruction also may offer the architect new tools for ensuring that the integrity of historic and prehistoric structures is protected.

Finally, the methods described in this paper are nondestructive or minimally harmful to architectural materials. The techniques are attractive, of course, because one method of conservation especially necessary in archeology is the hands-off approach.

REFERENCES

1. Borchers, P.E. 1977. *Photogrammetric Recording of Cultural Resources.* National Park Service: Washington, D.C.

2. Lyons, T.R., and T.E. Avery. 1977. *Remote Sensing: A Handbook for Archeologists and Cultural Resource Managers.* National Park Service: Washington, D.C.

3. Wolf, P.R. 1974. *Elements of Photogrammetry.* McGraw-Hill: New York.

Cleaning and Surface Repair

JOHN ASHURST

This paper discusses the importance of cleaning as part of the maintenance of historic building surfaces. Cleaning systems currently in use are reviewed with special reference to recent developments. Methods of surface repair and treatment of joints likely to be associated with the cleaning are described.

The Research and Technical Advisory Service of the Directorate of Ancient Monuments and Historic Buildings (DAMHB) is involved with the day-to-day maintenance and repair problems of historic stone buildings. It also prepares technical notes and checks cleaning and repair programs that may be supported by government grants. This activity involves contact not just with a wide range of buildings and surfaces, but with an ever wider range of skills, abilities, and available resources. Stemming from this involvement, a need has been established for more practical guidance on techniques of cleaning and repair that relate realistically to these frequently limited resources.

The work currently being carried out in conservation laboratories is providing essential knowledge both on the behavior of building and decorative stones and on the potential and limitations of various techniques of maintenance. Much of this work relates, properly, to the cleaning and conservation of stone surfaces that have high intrinsic value, and the major practical benefit may therefore be to the con-

---

John Ashurst *is Research Architect at the Directorate of Ancient Monuments and Historic Buildings, Department of Environment, United Kingdom.*

servator with special skills concerned with relatively small-scale surfaces.

Some owners of stone buildings, some architects, and some masonry contractors may decide that increasingly selective treatments and increasing knowledge of the complexity of stone deterioration presented and described in technical conservation papers are too far removed from the practical business of dealing with large and possibly heavily soiled surface areas in their ownership or care. Thus they may turn instead to unsuitable known methods or to new and untried services and materials on the assurance of technical trade literature that may seem refreshingly uncomplicated and does not "confuse with facts."

There is a gap—although happily it tends to be narrowing—between conservation and commercial specification This gap needs to be bridged. Recognition of different standards leads to good advice being given and being heeded. For instance, the conservation of stone objects of value will exclude much sandblasting, alkali and acid cleaning, and crude washing as wholly inappropriate. Still, these tools of the repair and cleaning industry are in everyday use and will continue in use. Such cleaning systems need upgrading by relevant and sensible specification, rather than by outlawing, which only serves to alienate site practice from good recommendations altogether. In this way, techniques of cleaning, such as clay poulticing, solvent packs and creams, ultrasonic descaling, precision blasting, and consolidation techniques using polyester and acrylic mortars and different silane systems, may be seen as compatible with the proper care of, for instance, a medieval table tomb, while washing, abrasive water jetting, controlled grit blasting, or mild caustic or acid solvent systems will not necessarily be ruled out as suitable means of removing the dirt from nineteenth-century cottages.

This is not to say that poor standards of repair and cleaning, or damage to any building, should be acceptable. It may be wise to postpone any activity of this kind altogether in cases where available techniques and labor are known to be inadequate; or it may be determined that only a limited amount of cleaning should be carried out to enable essential repairs to be completed.

Ways in which commercial cleaning, treatment, and repair techniques may be improved and which require more dissemination are suggested below.

WASHING

Most of the problems associated with washing limestone and brickwork result from the large volume of water employed. This may cause staining, migration of salts with subsequent efflorescences, and damage to internal fixings, timber, and plaster. Large quantities of water may be avoided if a properly atomized system is used with mesh filters and small orifices, but a wet-fog condition is difficult to maintain in practice, even with a closely sheeted scaffold. Of course, the finest available sprays should be used, and the best control may be to regulate them by a clock. Simple clock mechanisms may be preset to allow the jets to operate for, say, 10 seconds, and then shut off for 4 minutes before rewetting. The sprays should not be directed straight onto the soiled surface, but allowed to play across it. As progressive softening takes place, the dirt should be eased away with small, nonferrous-wire or fine bristle brushes. Clock-operated systems are easy to set up, and the timing can be adjusted to the type of stone and the amounts and types of dirt. To avoid risk of damage by freezing, washing should be programmed for frost-free months.

MECHANICAL CLEANING

Most mechanical cleaning is drastic and may be seen almost as a method of redressing the stone surface. Flexible carborundum discs and small carborundum heads have made the technique more sophisticated, but most surfaces cleaned in this way exhibit scour marks and must be finished by hand rubbing, removing even more of the surface. The method is best excluded except where disfiguration by oil, grease, and paint, coupled with physical damage, requires a new surface to be formed. To take back a surface in this way requires considerable skill if it is to look good, and it should only be done as a last resort.

STEAM CLEANING

The use of the steam lance is much less common than it was in the 1930s. After that time it fell into disrepute, partly because of the residual damage left by caustic soda used as a water softener and partly because it was often no more effective than cold-water washing. Today steam is useful as a support method where greasy deposits are encountered on stone surfaces or sticky substances need to be removed.

AIR ABRASIVE TECHNIQUES ("SAND/GRIT BLASTING")

Air-abrasive methods probably have caused more damage to masonry surfaces in the past two decades than any others. The rapid manner in which dramatic cleaning effects can be achieved has unfortunately been a strong selling point. Since higher pressures and coarser abrasives produce ever greater and more dramatic results, the commercial temptation to misuse the system is obvious. Air-abrasive techniques are used on a large scale to clean sandstone, limestone, brick, and even terra-cotta and stucco. Despairing of control, some authorities have banned the system in their areas of administration. However, many contractors have demonstrated their ability to clean satisfactorily in this way by using small units and fine abrasives at low pressures. Of course, it would be nonsense to suggest that a building front should be cleaned using an air-abrasive pencil and aluminum oxide abrasive, but there is a wide range of equipment now available, both wet and dry, that gives the operator full control and full visual command of the cleaning operation.

Specification should prescribe the air-abrasive gun, the abrasive type and size, and the acceptable pressures. Supervision should be regular, including, if necessary, spot checks on air pressures with a hypodermic needle and a gauge. Sensible cleaning times should be ascertained in advance so that the temptation to speed up and save money is reduced. An agreed sample of cleaning should be on-site and constantly referred to. Pressures over 40 psi should not be accepted, and in some cases a lower limit must be set. Abrasives preferably should be nonsiliceous, although it must be pointed out that large sections of the cleaning industry use nothing but sands of various mesh sizes, and the dust risk is certainly as great from a sandstone surface that is being cleaned as from the abrasive. Scaffolds should be tightly sheeted to protect the public, windows and other openings carefully protected and sealed, and the operators fitted with filtered air-line helmets.

Air-abrasive techniques are primarily of interest in cleaning sandstone. They may also be used on heavily soiled limestone, but should not be used on brick or terra-cotta. The work should be finished by rinsing the building face with clean water, preferably using a low-pressure water lance with a fan jet to remove adherent dust.

CHEMICAL CLEANING

The nonscientific building owner or specification writer is most vulnerable when beset by the multitude of chemical cleaning materials and services that promise to solve all his problems. Indeed, two aspects of chemical cleaning—the absence of water saturation and abrasion—are undeniably attractive. The price that is sometimes paid, however, is the damage resulting from residual soluble salts and disfigurement from staining or formation of white silica "bloom." Most alkaline cleaners are based on sodium hydroxide, and most acid cleaners are based on hydrofluoric acid. Additions of surfactants and rust inhibitors and the use of complex or subtle trade names often confuse the potential user further. Since it is unlikely that either cleaner will cease to be used, and in some cases may genuinely be the best available solution to a problem, sensible controls should be encouraged. In particular, those who are to use chemical cleaners must be fully aware of the potential damage to themselves and the building and prepare both themselves and the building with appropriate protective clothing and sheeting. Dilution of chemicals must not be permitted on site, and appropriate first aid must be taught and understood. All work must be to a previously agreed standard, set by a sample on-site. Proper prewetting to reduce absorption must be carried out and the contact between the chemical (preferably in the form of a thixotropic paste) and the soiled surface kept to a minimum. Washing may be preceded by lifting off the thixotropic paste with a suitable wooden scraper. The washing itself, using a low-volume, fan-jet water lance, must be carried out scrupulously and systematically, paying particular attention to joints. The use of "neutralizing" solutions (i.e., the use of oxalic acid after caustic alklai cleaning) is sometimes recommended, but it is difficult to see that this really contributes anything, and it is another costly operation.

One of the common uses of a caustic alkali cleaner, combined with a detergent, is to degrease before cleaning by another method. Caustic alkali is used on limestone and brickwork, and dilute hydrofluoric acid is used commonly on sandstone and granite, some brickwork, and some terra-cotta.

If the risks to both personnel and buildings are properly understood and rigid control and discipline relevant to the method are exercised, satisfactory cleaning may be achieved and may be preferred to other methods. Of course, there is a temptation to turn away from any chemical cleaner, but since they are entrenched as systems there is strong reason for trying to keep the techniques as useful and as safe as possible.

POULTICE TECHNIQUES

In some instances poultice packs, familiar in the monument conservation world, may be used simply and economically by the cleaning contractor. There is no reason, for instance, why magnesium silicate clay packs should not be applied to areas of detail to assist in softening encrustation. However, caustic materials are sometimes included in a body of clay to break down oil or grease, and lime or whiting bodies have been used traditionally with ammonium chloride to remove copper staining, or with sodium citrate to remove iron stains. These measures may be alarming in the context of monument conservation, but if carefully applied, they may be acceptable on less important structures. Once again, the specification must demand proper protection of surrounding areas, careful prewetting, lifting off by spatula after minimum contact time, and thorough and careful rinsing. Clay packing after the use of these poultices is sometimes useful.

SURFACE TREATMENTS

Water repellent materials such as silicone resins and metallic stearates are commonly recommended and applied as dirt inhibitors after cleaning. While the retention of a cleaner appearance has been observed on some surfaces, compared with cleaned but untreated areas, this is offset by the patchy appearance that may develop after seven or eight years and by the initial expense. These treatments should not, of course, be applied to decaying surfaces or to areas likely to be subjected to major moisture movements, and if they are to be used, sufficient time must be allowed for the surface to become as dry as possible. This point is often overlooked when a contract is running late.

Although material with shallow penetration into the surface must not be thought of as a preservative, it may assist indirectly to extend the life of stones or bricks. For example, the sound external face of stone tracery may be treated back to the glass line to reduce the feed-through of salt-contaminated water to an internal drying face that is decaying. Brickwork may be treated with the appropriate class of silicone to prevent staining and decay resulting from washing off limestone dressings.

Lime-based treatments, although lately associated with the conservation of external limestone sculpture and carved architectural detail, may be usefully applied to more modest limestone facades that are suffering from the effects of atmospheric pollution. The simplest treatments, but the ones that most substantially change the appearance of a surface, are the traditional tallow-based lime wash, a lime wash based on hydraulic lime, or lime and pozzolanic cenospheres. These may be economical to apply, but must be renewed at four- or five-year intervals, depending on exposure. They have the disadvantage of increasingly blurring and distorting the detail of the building. Sometimes, too, lime washing is applied over dirty surfaces as a means of trying to lighten and improve the appearance of a facade quickly and cheaply, a practice that must be discouraged.

Better than lime washing is the system of either washing or cleaning by hot lime poultice, and then rubbing in finely sieved lime putty and sand bound with a small amount of casein (inhibiting organic growth with formalin)—the method developed by Baker at Wells. Such a treatment closes up the texture of the stone without a drastic change in appearance and can be worked in well with small repairs in matching mortar.

Although treatments of this kind have also been applied to sandstone, such practice should be generally discouraged because of its tendency to encourage decay. There are some difficult exceptions; when sandstone has been treated with lime over a period of many years, the best course may be to continue lime washing. In this situation it is essential that the lime coat be maintained intact and not allowed to deteriorate.

A further treatment of interest is the so-called silicate paint system developed in the 1890s by the Keim family in Augsburg as a "permanent paint" system, primarily for stucco, but also applied extensively to sandstone, limestone, and more recently to concrete. Surviving pigmented Keim treatments of 60 years are not uncommon. While it is acknowledged that in theory such treatments, using sodium and potassium silicate, carry a risk of forming soluble salts, the Keim track record is good. The light consolidation achieved is such that the treatment of large areas of low value may well be justified, especially when the alternative is replacement.

The use of silanes is discussed at some length in other papers and is not included here for that reason alone. The direct labor force of DAMHB and certain contractors have been trained in the use of the Building Research Establishment's system, "Brethane," and have achieved some excellent results. The results support the generally, though not exclusively, held view that this class of materials is the most promising of all the surface treatments, with a genuine claim to be described as a "preservative."

SURFACE REPAIR

The most common surface repair is the refilling of weathered-out joints. In the past this pointing operation has often been carried out by using unsuitable materials with strong hydraulic cement binders, frequently flush-filling over weathered and rounded arrises, thereby establishing water traps. Retention of the original joint width and the use of mortars no stronger than the stones in which they are to be placed is accepted as good practice, but too rarely carried out. Joint filling may be accompanied by local hand grouting or may be part of a major grouting operation.

Liquid grouts and the flushing operation that must precede grouting are notorious for introducing or moving concentrations of salts, especially when cement grouts are used. Low sulfate cements, limes, and fly ash are increasingly used, along with various additives, and are better in this respect, but the flushing operation is still a problem and cannot be dispensed with.

Massive masonry walls can, however, often be consolidated satisfactorily by gravity or low-pressure grouting in small sections. The important point to cover in specifications is that valuable insertions in the wall, such as monuments, should be isolated from grout flows by an impervious barrier, which will involve some drilling or cutting out and rebuilding. Such barriers may be epoxy mortars, pitch-extended epoxy, or bitumen-coated lead (provided there is no staining risk). Polyester mortar or acrylic resin and sand mortar may be used to rebuild and protect monuments against salt contamination from the surrounding structure.

Mortars for tamping and pointing historic masonry are normally lime/aggregate mixtures in the proportions 1:3, but may be gauged with small quantities of cement or finely ground pozzolanic material. A hydraulic lime may be used if early strength and early resistance to frost are required. Although the dangers of wet mixes based on hydraulic cements must be appreciated, a sensible compromise must be reached whereby adequate performance is obtained from the mortar without the risk that harmful quantities of sodium salts will be transferred into the masonry. A cement/plasticizer mix may be used in place of lime when pointing and filling sandstone joints, especially if lime has contributed to peripheral decay of the stones. In this way further contamination may be avoided.

Proper preparation and storage of mortar materials for weak lime mortar should be encouraged. Evidence suggests that lime kept as putty and mixed with the desired aggregates as long as possible before use results in better performance than lime and aggregates mixed dry and used immediately after adding water. Time, as well as thorough mixing, beating, and ramming, is needed to allow the aggregate particles to become thoroughly coated with binder. Of course, cement and other pozzolanic materials must only be added just before use.

SUMMARY

The above notes are suggestions for specifications for contractors who are to carry out work on historic buildings. The suggested measures are compatible with conservation practice without being impracticable or too demanding.

BIBLIOGRAPHY

[Prepared at the committee's request by Anne Grimmer, National Park Service.]

Anon. Cleaning external surfaces of buildings. *Building Research Station Digest* 113(1970).

Ashurst, John. *Cleaning Stone and Brick.* Technical Pamphlet 4. Society for the Protection of Ancient Buildings: London, 1977.

Ashurst, John, and Francis G. Dimes. *Stone in Building: Its Use and Potential Today.* The Architectural Press: London, 1977.

Clarke, B.L. Some recent research in cleaning external masonry in Great Britain. In *The Treatment of Stone.* Centro per la Conservazione delle Sculture all'Aperto: Bologna, 1972.

Clayton, Ian. Special feature: Stone cleaning. Why buildings should be washed. *Building Conservation* 3(3):20 (1981).

Hempel, K. Notes on the conservation of sculpture: Stone, marble, and terra-cotta. *Studies in Conservation* 13: (1968).

Lewin, S.Z., and Elizabeth J. Rock. Chemical considerations in the cleaning of stone and masonry. In *The Conservation of Stone* I (proceedings of the International Symposium, Bologna, June 19–21, 1975), ed. R. Rossi-Manaresi. Centro per la Conservazione delle Sculture all'Aperto: Bologna, 1976.

Mack, R.C. *Preservation Briefs: 1. The Cleaning and Waterproof Coating of Masonry Buildings.* National Park Service: Washington, D.C., 1975.

Stambolov, T. Notes on the removal of iron stains from calcareous stone *Studies in Conservation* 13:45–47 (1968).

Stambolov, T., and J.R.J. Van Asperen de Boer. *The Deterioration and Conservation of Porous Building Materials in Monuments. A Review of the Literature.* International Centre for Conservation: Rome, 1972.

Weiss, Norman R. Cleaning of Building Exteriors: Problems and procedures of dirt removal. *Technology and Conservation* 2(76): 8–13(1967).

Weiss, Norman R. *Exterior Cleaning of Historic Masonry Buildings.* draft report. National Park Service: Washington, D.C., 1975.

Preventive Maintenance in Historic Structures

NORMAN R. WEISS

This paper reviews the role of the construction industry in the maintenance of historic structures. Cyclical inspection and maintenance procedures are derived from an understanding of the behavior of the entire building. This approach demands familiarity with materials and systems not normally considered within the domain of the masonry contractors. Problems with the use of modern industrial maintenance products are discussed in terms of both performance and advertising claims, the latter often suggesting uses that can be destructive to weathered materials. Deemphasis of craft skills in contemporary construction is also considered as a factor in the increasing inability of building owners to contract for careful duplication of historic and functional details.

In recent years we have barely begun to shed the linguistic apparel that has kept building maintenance isolated from the scientific investigative process. While the term "conservator" has come to mean a skilled materials specialist trained in building science, architectural history, and project management, the words "maintenance man" still leave us with an image of little more than the faceless fellow on the wooden end of a mop.

In principle, competent building maintenance has the capacity to make a unique and outstanding contribution to the preservation of

Norman R. Weiss *is Assistant Professor of Architecture and Planning, Columbia University.*

historic structures. Since 1974 this potential has been recognized and encouraged through training efforts by the National Park Service, the National Trust for Historic Preservation, and the Association for Preservation Technology. Viewed in terms of the duration of interaction with the structure, maintenance emerges as the most significant component of the conservation process. Maintenance personnel form a permanent team whose work continues long after the restoration crew has completed its more glamorous assignment.

In this essay I would like to examine the difference between maintenance theory and practice—to attempt to bridge the wide gulf between concept and performance.

The logical premises upon which a maintenance plan can be constructed are few and rather simple. First among them is that periodic inspection is a necessary component of the maintenance process. Upon the frequency of this type of examination depends the ability to locate problematic conditions soon after their appearance.

Second, the nonlinearity of many weathering processes makes deterioration more readily correctable in its early stages. Thus, appropriate remedial actions based on periodic inspection can almost be preventive. Frequent operations at a relatively low level of intervention can be scheduled on the basis of prior experience with the stability of the materials present and with the observed patterns of building use.

When maintenance operations are performed in-house, the physical closeness of the staff to the problem is a third asset. The basic issue, once again, is swiftness of response, combined in this instance with a day-in day-out familiarity with the building's systems.

Finally, the record-keeping associated with maintenance administration is a valid means of trial-and-error learning and also provides important information to guide larger conservation efforts. Examination of data recorded over a significant period can provide evidence of the lifetime of building materials in actual service. Such evidence is often more reliable than that generated by simulating weathering via accelerated testing.

Unfortunately, many maintenance men have relatively little craft experience in traditional building trades and they have insufficient exposure to the new technology of architectural conservation. Training programs are necessary to impart even the most basic information on identifying materials and on conventional cleaning and repair procedures. Emphasis here must be on meetings, lectures, and demonstrations rather than on written texts—maintenance manuals, training pamplets, and regulation books are not popular reading in the employees' lunchroom. Perhaps because of this uncomfortable relationship of maintenance personnel with the written word, record keeping tends in reality to be erratic.

A further problem is frequent employee turnover in low-level positions, which undermines the effectiveness of educational programs. Employees in supervisory positions, while not necessarily more competent, have the benefit of seniority. They may, however, have been frequently transferred from site to site, having had little opportunity to learn the workings of any structure for which they have been responsible.

These comments, of course, have been directed toward predictable events in the lifetime of a building. When the unexpected occurs, maintenance personnel frequently avoid diagnosing the condition and seek instead to treat symptoms observed. Administrators wrongly place considerable emphasis on mopping up water and painting over stains, rather than on stopping leaks at their point of origin.

When the type or extent of required remedial action calls for the services of a mason contractor, the ability to respond quickly may disappear. Contracting can be a time-consuming task, especially with publicly-owned buildings that may require allocating a significant amount of staff time to the preparation of forms.

Similarly, contractural problems plague the in-house user of maintenance products. Technical-product literature—much of it laced liberally with quasiscientific gibberish—must often be interpreted by purchasing officers, who are generally unfamiliar with the maintenance process. Their primary responsibility is to determine the equivalency of apparently similar products that may be cheaper, made locally, or distributed by a more reliable source. To accomplish this task on any basis of scientific fact is frequently impossible.

Amidst this frustration lies another, somewhat concealed set of problems that affects the construction industry at large. In the past few decades the role of the U.S. construction worker has changed. Today he is less of a craftsman and more of a heavy equipment operator. In some instances, he is merely an assembler of prefabricated parts. It hardly seems surprising, then, that today's average mason knows nothing about the tuckpointing of eighteenth-century brickwork.

A number of other problems seem less related to the so-called progress of contemporary construction technology than to the overall organization of the industry. Trade specialization makes it difficult for craftsman and contractor alike to understand the reciprocal relationships between stone and other building materials—wood, metals, paints, and so on. These relationships affect the pattern of the normal aging process of structures; they play a major part in the overall failure of a substantial percentage of large-scale masonry conservation efforts. For in-house work, there is frequently little understanding of the role played by such common operations as floor cleaning and deicing in the decay process.

Materials manufacturers, another part of the industry, seem at times to work at cross-purposes to us all. Their products are compositionally remote from the analytical reagent-grade materials that we test in our laboratories. They are, by and large, cheap, although a 55-gallon drum of anything is expensive these days. To survive in the marketplace, a good product must still be affordable.

Even when considerations of cost can be set aside, performance data may be irrelevant in product selection. Many of the products offered for preservation/maintenance have been developed for and tested on new construction systems. In cleaners, coatings, and admixtures, this is where the profits are.

There is a curious paradox in all of this, in that the attitude of organized labor toward maintenance workers has only recently started to change. Traces of the poor-cousin status persist, despite the fact that building maintenance is a rich source of jobs. For instance, construction unemployment has risen to 13 percent, twice the national rate for several other industries. But in New York City, Local 66—the Pointers, Cleaners, and Caulkers local—has 100 percent of its men working. The number of new housing starts, used as an indicator of our national economic health, is manipulated by fiscal planners, creating a stop-and-start situation. But building maintenance, by its very nature, goes on.

I mention Local 66 because it is of particular interest to me. In the fall of 1977, officials of this local, representing the International Union of Bricklayers and Allied Craftsmen, agreed to cooperate wih a training project called RESTORE. This project was developed by the Municipal Art Society, a not-for-profit organization with a board of socially and culturally prominent New Yorkers and a young and active staff. This curious amalgam of bricklayers and preservationists was created to disseminate building-maintenance restoration know-how to journeyman masons, talented apprentices, and mason contractors.

Several of the participants in this conference have appeared as guest speakers for RESTORE, sharing with me the presentation of 50 classroom hours of lectures and nearly 30 hours of laboratory and workshop demonstrations each year. We are looking forward to the inception this fall of a second program of shorter, more intensive courses in both masonry and plasterwork conservation. These will be offered in a number of cities throughout the country.

RESTORE has succeded in part because it responds to a need for information that can be put directly into the mechanic's hands. That this need is real is supported by recent national figures showing that approximately 85 percent of all work on existing buildings is done within a simple contractor-owner relationship—without a restoration architect, architectural conservator, or structural engineer.

Other successes are possible. Preventive maintenance, however, is grossly underfunded. This is as true for relatively new buildings as it is for historic ones. And it is true despite the fact that 35 percent of the construction dollars spent in 1979 were for preservation/maintenance. Look carefully at buildings in your own neighborhood and you will see that economic pressures have created a brutal emphasis on the functionality of repair work.

Yet I am certain that the construction industry can refine its methods if we can somehow instill in today's tradesmen an antiquarian affection for old buildings. We need to balance the "fix-it-up-cheap" demands of the marketplace with a genuine personal concern for a different kind of maintenance, the maintenance of visual and historic quality.

BIBLIOGRAPHY

[Prepared at the committee's request by Anne Grimmer, National Park Service.]

Architectural Resources Group. *Checklist for Building Code Compliance and Building Inspection Checklist.* Compiled by Bruce Judd (203 Columbus Avenue, San Francisco, CA 94133), 1980.

Chamber, J. Henry. *Cyclical Maintenance for Historic Buildings.* National Park Service: Washington, D.C., 1976.

Council for the Care of Churches. *How to Look After Your Church.* London, 1970.

Davey, Andy, Bob Heath, Desmond Hodges, Roy Milne, and Mandy Palmer. *The Care and Conservation of Georgian Houses: A Maintenance Manual.* The Architectural Press with the Edinburgh New Town Conservation Committee: London, 1980.

Gilder, Cornelia Brooke. *Property Owner's Guide to the Maintenance and Repair of Stone Buildings.* Technical Series/No. 5. The Preservation League of New York State: Albany, 1977.

Holmstrom, Ingmar, and Christina Sandstrom. *Maintenance of Old Buildings Preservation from the Technical and Antiquarian Standpoint.* National Swedish Building Research Document D10. National Swedish Institute for Building Research: Stockholm, 1975.

Huhn, Tom. *Directory of Training Programs and Information Resources on Restoration and Preservation Building Trades and Crafts.* Fact Sheet. National Trust for Historic Preservation: Washington, D.C., 1981.

Insall, Donald. *The Care of Old Buildings Today: A Practical Guide.* The Architectural Press: London, 1972.

Mack, Robert C. *Preservation Briefs: 1. The Cleaning and Waterproof Coating of Masonry Buildings.* National Park Service: Washington, D.C., 1975.

Melville, Ian A., and Ian A. Gordon. *The Repair and Maintenance of Houses*. The Estates Gazette Limited: London, 1973.

Pierpont, Robert N. *A Primer Preservation for the Property Owner*. The Preservation League of New York State: Albany, 1978.

Stahl Associates, Inc. *Maintenance, Repair and Alteration of Historic Buildings*. HP Document No. 1. General Services Administration: Washington, D.C., 1981.

Weiss, Norman R. Cleaning of building exteriors: problems and procedures of dirt removal. *Technology and Conservation* 2(76):8–13(1976).

Weiss, Norman R. *Exterior Cleaning of Historic Masonry Buildings*. Draft Report. National Park Service: Washington, D.C., 1975.

Stone-Consolidating Materials: A Status Report

JAMES R. CLIFTON and GEOFFREY J. C. FROHNSDORFF

Mechanisms by which stone consolidants function are outlined. Evaluation of stone consolidants usually requires both laboratory and field tests to determine their initial and long-term performances. ASTM Standard E 632, Recommended Practice for Development of Accelerated Short-Term Tests for Prediction of the Service Life of Building Materials and Components, can be used to provide guidance on the test program.

Materials that have been investigated as stone consolidants are reviewed. They fall into four main groups: inorganic materials, alkoxysilanes, synthetic organic polymers, and waxes. Epoxies, acrylics, and alkoxysilanes are the most commonly used consolidants, but no consolidant can be considered completely satisfactory and able to meet all the desired performance requirements.

Building stones may lose their integrity (i.e., decay) as a result of weathering.[1] Loss of material from the exposed surfaces of stone masonry units and the reduction in compressive strength and other mechanical properties of the units usually proceed slowly. However, such changes

James R. Clifton is Group Leader, Inorganic Materials, Building Materials Division, Center for Building Technology, National Bureau of Standards, Washington, D.C. Geoffrey J. C. Frohnsdorff is Chief, Building Materials Division, Center for Building Technology, National Bureau of Standards, Washington, D.C.

The authors wish to acknowledge the encouragement of Hugh Miller, Chief Historical Architect of the National Park Service, and Henry Judd, former Chief Historical Architect of the National Park Service, both of whom provided information valuable to this review.

are likely to be significant in the case of structures that we hope to preserve for many future generations. Problems caused by loss of integrity are well known for masonry units consisting of porous sedimentary rocks such as sandstone and limestone. The decay is generally believed to result from dissolution of the material cementing the grains of stone together, or from disruption of the intergranular bonds from increased tensile stresses caused by such processes as salt crystallization and thermal expansion.

Intergranular bonds in a highly porous stone can be represented schematically, as in Figure 1. The drawings represent a magnified cross

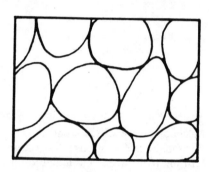

a. The original stone; alternatively, decayed stone treated with a consolidant which accumulates at the contact points to restore bonds between grains.

b. Stone treated with a consolidant which provides a uniform thin coating on the grains and bonds them at the contact points.

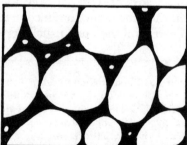

c. Stone treated with a consolidant which almost fills the pores.

FIGURE 1   Schematic diagrams representing magnified cross-sections through a porous stone treated with consolidants.

section through limestone or sandstone. (In reality, of course, a planar section through a stone would be unlikely to intersect so many contact points between adjacent grains.) If a sufficiently large proportion of the cemented bonds at the contact points is broken within any volume of the stone, the integrity of the stone in that volume will be lost. This usually happens close to the surface. If it is necessary to restore the integrity of decayed stone, the stone must be treated with a material that will effectively restore the bonds between adjacent grains.[2] Materials used for treating stone to restore integrity are termed stone consolidants. While there is evidence that decayed stones can be consolidated, at least in the short term, knowledge of stone consolidation is not at a level where the performance of consolidants over many years can be confidently predicted and the treatments guaranteed not to harm the stone.

To provide perspective on stone consolidants, reference will be made to Figure 1. Figure 1a represents the original stone with cemented bonds between the grains; it could also represent what might be achieved if, following the breaking of intergranular bonds, a stone consolidant could be made to accumulate in the contact areas to reestablish intergranular adhesion. Figure 1b represents a decayed stone which has been treated so as to produce a thin coating of consolidant covering the surface of each grain and bonding the grains together at the contact points. Figure 1c represents a decayed stone which has been treated with a consolidant that almost completely fills the pores of the stone, leaving only relatively small voids or pores within the consolidant. The diagrams show, in a simplistic way, how stone consolidants may affect the microstructure of a treated stone and influence the properties of the surfaces and interfaces within it.

Treated stones are composite materials, and their properties reflect the properties of their individual ingredients, the interactions among them, and their spatial distributions. Since the properties and long-term performance of such composites cannot be satisfactorily predicted, the selection of consolidating materials and treatments must usually be based on laboratory and field tests of treated stones. To aid the necessarily complicated development of durability tests, American Society for Testing and Materials (ASTM) Subcommittee E 6.22 recently established standard E 632, Recommended Practice for Development of Accelerated Short-Term Tests for Prediction of the Service Life of Building Materials and Components.[3,4] Figure 2, which is taken from ASTM E 632, summarizes the recommended practice. It outlines an approach to follow if the durability of a treated stone is to be evaluated in a rational way.

Stone consolidants are applied as liquids but, to be effective, they

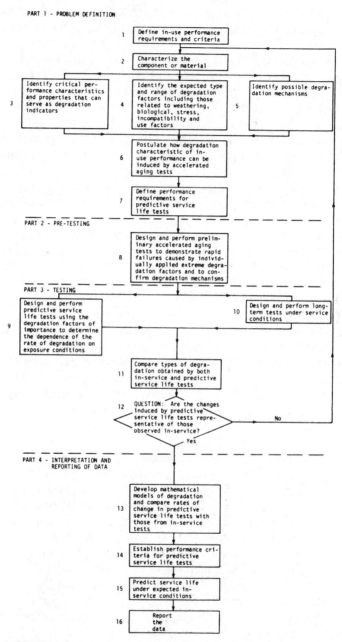

FIGURE 2 Steps in the recommended practice for developing predictive service life tests, ASTM Standard E 632.

must cause a solid material to be laid down in the pores of the stone. The initial properties of a stone consolidant depend on many factors. The penetration of the consolidant into the stone and its distribution within the stone depend on the structure of the stone, the viscosity and surface tension of the liquid, and the contact angle of the liquid against surfaces within the stone.[5,6] Consolidation may result from solidification of the liquid within the pores, as by polymerization of a monomer or cooling of a molten solid, or from evaporation of a volatile solvent from a solution of a resin or other solid material. It may also result from nucleation and growth of relatively small quantities of solid from the liquid phase. Examples of consolidants depending on each of these mechanisms are given below under Stone Consolidants.

The durability of a consolidated stone depends in part on the durability of the consolidant in the environment encountered in service. It may also reflect the fact that the distribution of stresses within a treated stone may differ from that in the untreated stone because of the changes in microstructure and the characteristics of the exposed surfaces. A list of degradative factors that should be considered in evaluating the durability of any material is given in Table 1, and a matrix to aid the identification of possible degradation mechanisms in individual phases and at interfaces between phases is given in Figure 3.[3,4]

With this general discussion as background, the range of stone consolidants that have been used, or proposed for use, will now be reviewed. The review is based on a previous paper by James R. Clifton, which gives a more extensive bibliography.[2]

STONE CONSOLIDANTS

In this review, stone consolidants are divided into four main groups: inorganic materials, alkoxysilanes, synthetic organic polymers, and waxes. Considerations of their performance are based on generally applicable requirements, discussed in Clifton's paper.

Inorganic Materials

Inorganic stone consolidants were used extensively during the nineteenth century and are still used occasionally. Most inorganic consolidants produce an insoluble phase within the voids and pores of a stone, either by precipitation of a salt from the liquid or by chemical reaction of the liquid with the stone. It has been suggested that the development of a new phase similar in composition to the matrix of a stone will

TABLE 1   Degradation Factors Affecting the Service Life of Building Components and Materials

---

Weathering Factors
  Radiation
    Solar
    Nuclear
    Thermal
  Temperature
    Elevated
    Depressed
    Cycles
  Water
    Solids (snow, ice)
    Liquid (rain, condensation, standing water)
    Vapor (such as high relative humidity)
  Normal air contaminants
    Oxygen and ozone
    Carbon dioxide
  Air contaminants
    Gases (such as oxides of nitrogen and sulfur)
    Mists (such as aerosols, salt, acids, and alkalies dissolved in water)
    Particulates (such as sand, dust, dirt)
  Freeze–thaw cycles
  Wind

Biological Factors
  Microorganisms
  Fungi
  Bacteria

Stress Factors
  Stress, sustained
  Stress, periodic
  Stress, random
    Physical action of water, as rain, hail, sleet, and snow
    Physical action of wind
    Combination of physical action of water and wind
    Movement due to other factors, such as settlement or vehicles

Incompatability Factors
  Chemical
  Physical

Use Factors
  Design of system
  Installation and maintenance procedures
  Normal wear and tear
  Abuse by the user

---

be effective in binding together the grains of deteriorated stone. For example, consolidants that result in the formation of a siliceous phase should be used to consolidate sandstone, and calcium carbonate or barium carbonate should be used to consolidate calcareous stones such as limestone. In practice, however, little concern seems to be given to chemical compatibility between the consolidants and the stone.

Little success has been achieved in consolidating stone with inorganic materials, and in some cases their use has greatly accelerated decay.[7-9] Some of the reasons given for the poor performance of inorganic consolidants are their tendencies to produce shallow, hard crusts,[7,10,11] the formation of soluble salts as reaction by-products,[7,10,12,13,60,94] growth of precipitated crystals,[8] and the questionable ability of some of them to bind particles of stone together.[14,15] Of these, the most difficult problem to overcome is the formation of shallow, hard surface layers by inorganic consolidants because of their poor penetration abilities. Precipitation processes are often so rapid that precipitates are formed before the inorganic chemicals can appreciably penetrate the stone. Precipitation from homogeneous solutions has been used to obtain deeper penetration of stone by some inorganic consolidants. This method is discussed below under Alkaline Earth Hydroxides.

Siliceous Consolidants

Siliceous consolidants are materials that have been used to consolidate sandstone and limestone through the formation of silica or insoluble silicates.

Alkali Silicates Both nonstoichiometric dispersions of silica in sodium hydroxide and soluble alkali silicates have been used to preserve and consolidate stone. When dispersions of silica in sodium hydroxide solutions are applied to a stone, silica is deposited.[7,16] If sodium hydroxide is not removed by washing, it can react with carbon dioxide or sulfur trioxide to form sodium carbonate or sodium sulfate, respectively. These salts may cause unsightly efflorescence and salt crystallization damage. In addition, it seems that sodium hydroxide can react with the constituents of some stones, thereby accelerating deterioration.[7]

Silica can be precipitated by the reaction between sodium silicate or potassium silicate and acids such as hydrochloric, arsenic, and carbonic acids.[7,16,17] However, these reactions result in the formation of soluble salts such as sodium chloride and sodium arsenate. If the so-

294

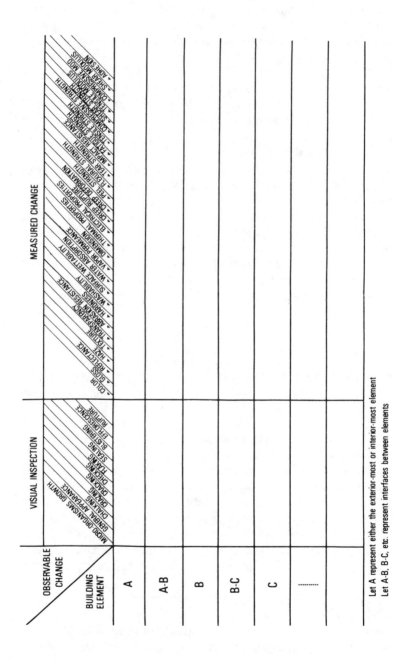

FIGURE 3 Example of a matrix for identifying observable changes of building components and materials (from ASTM Standard E 632).

dium silicate–arsenic acid mixture is used to consolidate limestone, crystalline calcium arsenate can be produced by a reaction between calcium carbonate and arsenic acid. The crystalline calcium arsenate appears to damage limestone by anisotropic crystal growth.[7]

Insoluble silicates have been precipitated in stone by alternate treatments of sodium silicate and salts such as calcium chloride[7,15,16,18] and zinc carbonate.[19,97] The colloidal silicates that are first produced eventually crystallize, while soluble salts are produced as by-products.[7] Also produced are impervious surface layers that trap underlying water.[20] Apparently the silicates precipitate relatively rapidly and are deposited near the surfaces of the treated stones.

Even with all the problems associated with the use of alkali silicates, they are still occasionally applied.[21] Recently, the successful use of soluble silicates was reported.[22] However, present evidence suggests that alkali silicates should not be used as stone consolidants.

Silicofluorides Both hydrofluosilicic acid and soluble silicofluorides have been used to preserve and consolidate stone. Hydrofluosilicic acid should not be used on limestone because it reacts vigorously with calcium carbonate to form crystalline calcium silicofluoride, carbonic acid, and carbonate salts.[18] The reaction occurs upon contact of the acid with the limestone, producing a shallow crust with little consolidating value. Hydrofluosilicic acid reacts more slowly with siliceous-based sandstones to form a cementitious material, but again only the surface is hardened. Hydrofluosilicic acid has a tendency to discolor both limestones and sandstones, especially if they contain iron.[7] Many soluble silicofluorides, such as those of magnesium, zinc, and aluminum, have been applied to limestone. Resulting products are silica, insoluble fluoride salts, and carbon dioxide, which are formed near the surface of the limestone. Therefore, only the surface is hardened, and eventually it exfoliates.[11,23,24] Soluble silicofluorides also react with calcareous sandstones, and again only a hardened surface is obtained. Further, soluble salts are formed when both limestone and calcareous sandstone are treated with silicofluorides.[10] These soluble salts have caused damage through salt recrystallization processes.[12] Penkala recently carried out a systematic study of several stone treatments and also found that fluorosilicates were not effective consolidants.[25]

Alkaline Earth Hydroxides

Calcium Hydroxide Aqueous solutions of calcium hydroxide (its saturated solution is often called limewater) have been used for many centuries to protect and consolidate limestone.[26] Calcium hydroxide itself does not appear to consolidate stone, but in solution or in a wet state it reacts with atmospheric carbon dioxide to form insoluble calcium carbonate, which may bind particles of calcareous stones together. The solublility of calcium hydroxide is only about 1 g per liter at room temperature;[27] therefore, repeated applications are necessary to produce sufficient calcium carbonate to consolidate stone. Furthermore, unless very dilute solutions are used, only the calcium hydroxide deposited near the surface of a stone is carbonated. This happens if the dense calcium carbonate formed at the surface fills the pores and voids in the stone and severely impedes the migration of carbon dioxide through the treated surface to the interior. The newly produced calcium carbonate is susceptible to the same deterioration processes as the calcareous stone. For example, it can react with sulfur trioxide to form calcium sulfate, which is relatively soluble compared to calcium carbonate. Therefore, the treated stone may not be protected against further weathering. However, it may eventually gain the authentic appearance of the weathering stone.

Conflicting opinions have been given of the effectiveness of the calcium hydroxide process. Some conservators have felt that while treatment with calcium hydroxide causes no harm, little permanent consolidation is obtained,[7,101] while others have recommended the use of limewater to protect limestones from weathering and to consolidate them.[8,26,28,29] The effectiveness of freshly prepared slaked lime (calcium oxide mixed with water) in consolidating statues at the Wells Cathedral in England is being investigated by Baker.[30] He applies 38 mm thick layers of slaked lime to statues and removes the layers several weeks later. Some consolidation appears to occur. Apparently, repeated treatment with limewater and slaked lime can gradually consolidate limestone, but such processes are economically feasible only for small objects.[49]

Strontium and Barium Hydroxides   Like calcium hyroxide, strontium and barium hydroxides will react with carbon dioxide to form insoluble carbonates, but again, only the hydroxide near the surface of a stone is carbonated. The carbonate may subsequently be converted to sulfate by interaction with atmospheric sulfur oxides. However, unlike calcium sulfate, the strontium and barium sulfates that may be formed are insoluble. Thus the application of strontium and barium hydroxides may reduce the weathering of stone exposed to environments polluted by sulfur oxides.

The early work on the use of barium hydroxide to preserve stone was performed by Church.[31,32,33] Initially, excellent results appeared to be obtained. However, only surface hardening occurred, and the barium carbonate or barium sulfate layer eventually exfoliated.[7,8,11,20] The exfoliation problem has been attributed not only to the formation of a dense, impervious surface layer, but also to anisotropic crystal growth of barium carbonate and barium sulfate.[7,8]

Lewin and Sayre have developed methods intended to precipitate barium carbonate and barium sulfate deeply within a stone.[34,35] These methods are based on precipitation from homogeneous solution.[36] In this process, the material to be precipitated and the precipitating chemicals are present in the same solution. For example, barium carbonate is precipitated from an aqueous solution of barium hydroxide and urea.[34,37] The urea hydrolyzes slowly, at a rate dependent on the pH, to produce ammonium carbonate (or ammonia and carbon dioxide) in the solution. This causes the pH to rise and the hydrolysis to accelerate. At the same time, the carbonate reacts with the barium ions in solution to precipitate barium carbonate. The reaction rate can be controlled so the precipitate forms days after a stone is treated. The slow formation of barium carbonate is reported to give a crystalline-solid solution with the calcite crystals of calcareous stone. Barium sulfate can be precipitated in a stone by an analogous method. An aqueous solution of a barium monoester of sulfuric acid hydrolyzes slowly when a base is added, releasing barium and sulfate ions.[36]

The precipitation of barium carbonate and barium sulfate from homogeneous solution is a promising approach. To date, however, only experimental testing has been carried out, and little is known of the long-term consolidating effectiveness of this approach. Warnes and Marsh have both suggested that crystalline inorganic precipitates, such as barium carbonate and sulfate, do not have long-term consolidating value.[7,8] They have also commented that the precipitates of barium carbonate and barium sulfate have larger volumes than calcite and appear to exhibit anisotropic crystal growth. It should not be assumed that deteriorated stone will have sufficient empty volume to accommodate these precipitates. Therefore, until more is known of the long-term effects of barium carbonate and barium sulfate on the durability of stone, they should be regarded as experimental materials and should not be applied to important historic structures.

Other Inorganic Consolidants

Many other inorganic materials have been used in attempts to preserve or consolidate stone. They include zinc and aluminum stearates,[7,8,25,38] aluminum sulfate,[7,8,36] phosphoric acid,[8] phosphates,[8] and hydrofluoric acid.[32] Hydrofluoric acid appears to have a consolidating effect because it removes deteriorated stone, thereby leaving a sound surface. A saturated aqueous solution of calcium sulfate has been used recently to consolidate stone consisting of a conglomerate of microfossils cemented by gypsum.[39,95]

Alkoxysilanes

Uses and Developments

Alkoxysilanes are regarded by many stone conservators as being among the most promising consolidating materials for siliceous sandstones.[29,40–47,101] The feasibility of using alkoxysilanes to consolidate calcareous stone is also being studied.[48,49] The main reasons that alkoxysilanes are considered promising are their abilities to penetrate deeply into porous stone and the fact that their rates of polymerization can be adjusted to permit deep penetration.[29,40,42,43,45,50] In addition, they polymerize to produce materials that may be similar to the binder in siliceous sandstone.

The use of alkoxysilanes for consolidating stone is not a recent development. For example, A. P. Laurie received a patent in 1925 for producing such a material to be used for stone consolidation.[51] Other early researchers on the use of alkoxysilanes to consolidate stone are Cogan and Setterstrom.[52,53] Alkoxysilanes have been commonly used since 1960 in Germany.[21] And recently, a promising alkoxysilane consolidating material called Brethane has been developed at the United Kingdom Building Research Establishment.[45]

Alkoxysilane Chemistry

Alkoxysilanes are a family of monomeric molecules that react with water to form either silica or alkylpolysiloxanes. Three alkoxysilanes are commonly used to consolidate stone: tetraethoxysilane, triethoxymethylsilane, and trimethoxymethylsilane.[42] Tetraethoxysilane is an example of a silicic acid ester.[43] Polymerization is initiated by hydrolysis:

$$-\overset{|}{\underset{|}{Si}}-OR + H_2O \xrightarrow{\text{catalyst}} -\overset{|}{\underset{|}{Si}}-OH + ROH$$

Then polymerization commences:

$$-\overset{|}{\underset{|}{Si}}-OH \;+\; -\overset{|}{\underset{|}{Si}}-OR' \;\longrightarrow\; -\overset{|}{\underset{|}{Si}}-O-\overset{|}{\underset{|}{Si}}- \;+\; R'OH$$

where R = CH_3 (methyl) or C_2H_5 (ethyl), and R' = H, CH_3, or C_2H_5.

Polymerization continues until all the alkoxy groups have been liberated and either an alkylpolysiloxane or silica is produced. Silica is produced by the polymerization of a silicic acid ester. An alkylpolysiloxane is formed by the polymerization of other types of alkoxysilanes. An acidic catalyst (e.g., hydrochloric acid) is used to increase the rate of hydrolysis. The alkoxysilanes are diluted with solvents to reduce their viscosities. Thus, their reaction rates and depths of penetration into stone can be controlled. It is claimed that their consolidating ability can be increased by using a mixture of alkoxysilanes.[43]

Some confusion appears in the literature regarding the differences between silicon esters, silicones, and alkoxysilanes. Silicon esters are partially polymerized alkoxysilanes that still have ester groups attached to silicon. Silicones are polymerized alkoxysilanes that are dissolved in organic solvents and used as water repellents.[43]

Performance of Alkoxysilanes

Weber and Price have observed that alkoxysilanes can usually penetrate porous stones to a depth of 20 to 25 mm.[44,50] The newly developed Brethane has been reported to penetrate as deeply as 50 mm.[45] No noticeable polymerization occurs with Brethane for at least three hours after it is mixed with a solvent and catalysts.[42,50]

Marschner reported that alkoxysilanes improved the resistance of sandstone to sodium sulfate crystallization.[54,98] However, she also observed that their performance varied from sandstone to sandstone and also depended on the compatibility between the solvent and the specific stone being treated. Similar findings were reported by Moncrieff, who studied the consolidation of marble.[48] Snethlage and Klemm observed in a scanning-electron-microscope analysis of impregnated sandstone that a polymerized alkoxysilane appeared to fill the space between sandstone grains and form a continuous coating.[55] However, polymerized alkoxysilanes are reported to have little effect on the passage of moisture in stone and the frost resistance of stone.[40,43,48,50] Some slight changes in the color of treated stone have been observed.[56,57] For example, statues on Wells Cathedral have become a duller grey following treatment with an alkoxysilane. Further, a treated stone panel on the cathedral has acquired a slightly more orange tone than adjacent, untreated panels.

Once a section of stone is treated with alkoxysilane, it will probably weather differently from the untreated stone. Thus, unless most of the visible parts of a structure are similarly treated, the contrast between the treated and untreated stone could become very noticeable.

Strength improvements of around 20 percent have been reported for sandstone specimens impregnated with alkoxysilanes.[40,43] The ability of alkoxysilanes to consolidate deteriorated stone in the field, however, has not been demonstrated unequivocally. Further, it appears that the performance of alkoxysilanes varies from stone to stone.

Even if alkoxysilanes are found to be effective consolidants, their high cost will probably limit their use to statues and smaller-sized stone objects.[38,45]

Synthetic Organic Polymer Systems

Two main types of synthetic organic polymer systems are used to consolidate stone. In the first, polymers dissolved in appropriate solvents are applied to stone. They are deposited within the voids and pores as the solvent evaporates. In the second, monomers, either pure or dissolved in a solvent, are polymerized within the voids and pores of a stone. Viscous monomers are diluted with solvents so that deep penetration can be achieved.[58] However, solvents that evaporate rapidly (many common organic solvents) have been found to draw organic consolidants back to the surface of a stone, resulting in the formation of hard, impervious surface crusts.[58,59] Munnikendam recommends the selection of organic consolidants whose solidification does not depend on evaporation of solvents.[41]

Among synthetic polymer systems, both thermoplastics and thermosets have been used to consolidate stone. A thermoplastic is a material that can be reversibly melted by the application of heat without significant change in properties. Examples of thermoplastics are poly(vinyl chloride), polyethylene, nylon, polystyrene, and poly(methyl methacrylate). A thermoset is a material that can be formed into a permanent shape and hardened by the application of heat and, once formed, cannot be remelted or reformed. Polyester, epoxy, and polyurethane are examples of thermosets. Methyl methacrylate can be converted into a thermoset by copolymerization with a cross-linking material.

The use of synthetic organic polymer systems to consolidate stone is a recent development, dating back to the early 1960s. Therefore, little is known of the long-term performance of these materials. Some organic consolidants have been found to improve the mechanical properties of deteriorated stone significantly. Many organic polymers are susceptible to degradation by oxygen and ultraviolet radiation, but this should only affect the materials on the surface of a treated stone.[60] Riederer reported that the surfaces of some structures in Germany that had been consolidated with organic polymers in 1965 had exhibited deep channel erosion by 1975.[21] Apparently water gradually eroded the consolidated surface and, once the protective surface layer was pierced, the untreated stone was eroded rapidly.

Acrylic Polymers

Methyl methacrylate and, to a lesser extent, butyl methacrylate have been used to consolidate concrete[61,62] and stone.[50] These monomers can be applied solvent-free to porous solids and can be polymerized in situ. An excellent source of information on their polymerization, as well as on polymer-impregnated concrete, is the report by Kukacka et al.[61] Methyl methacrylate has been polymerized into poly(methyl methacrylate) by heating with an initiator, by gamma radiation, and at ambient temperature by a combination of promoters and initiators.[61,63] For thermal polymerization, the chemical initiator (catalyst) 2,2'-azobis(isobutyronitrile) has been found to be effective.[64] Heating blankets could be used to polymerize thermally methyl methacrylate or other monomers applied to a stone structure. Polymerization by radiation must usually be carried out in special chambers because of the radiation hazards. Chemical promoters convert initiators into free radicals at ambient temperatures, and the free radicals induce the polymerization of methyl methacrylate. Munnikendam used N,N-dimethyl-p-toluidine to decompose benzoyl peroxide into free radicals.[59] He found, however, that oxygen inhibited the subsequent polymerization reaction of methyl methacrylate. Better success probably could be achieved by using 2,2'-azobis(isobutyronitrile) as the initiator.[64]

Where deep impregnation and complete polymerization was achieved, methyl methacrylate and other acrylates have been shown to improve substantially the mechanical properties and durability of porous materials such as concrete.[61] However, incomplete impregnation with acrylates may result in the formation of a distinct, probably undesirable, interface between treated and untreated stone.[54]

As shown by their stress-strain curves, concretes impregnated with acrylic-based polymers are classified as brittle materials.[61,64,65] Stone consolidated with methyl methacrylate and other acrylics can be expected to exhibit similar brittle behavior.

Methyl methacrylate can harden the surface of a stone and effectively consolidate the stone if both deep penetration and complete polymerization are achieved. However, as is the case with alkoxysilanes, stone impregnated with methyl methacrylate will probably weather differently from untreated stone. In addition, erosion through the treated stone could contribute to the development of an unsightly appearance.[21]

Acrylic Copolymers

Copolymers are produced by joining two or more different monomers in a polymer chain.[66] A commercially available acrylic copolymer used for stone consolidation is produced from ethyl methacrylate and methyl acrylate.[55,67] Other acrylic copolymers that have been studied for stone conservation include copolymers of acrylics and fluorocarbons [68,69,99] and of acrylics and silicon esters.[41,55] The acrylic copolymers are dissolved in organic solvents and then applied to stone. As discussed earlier, unless very dilute solutions are applied, solvent evaporation will tend to draw the acrylic copolymers back to the surface of a stone. Then, even if diluted to the lowest concentration that will give some consolidation, their solutions may still have high viscosities, which will impede their penetration.

Vinyl Polymers

Several vinyl polymers have been studied or used for preservation and consolidation of stone. They include poly(vinyl chloride),[70,71,96] chlorinated poly(vinyl chloride),[71] and poly(vinyl acetate).[67,70,71,72] These polymers are dissolved in organic solvents and then applied to stone. Photochemical processes could release chlorine from these chlorine-containing polymers, which could damage stone.[71] Poly(vinyl acetate) has been found to produce a glossy stone surface.[71] If vinyl polymers are not sufficiently diluted and carefully applied, their use undoubtedly will result in the formation of impervious layers which could entrap moisture and salts within the stone.[67]

Epoxies

An epoxy consists of an epoxy resin and a curing agent, which is actually a polymerization agent. Mixing the epoxy resin with the curing agent converts it into a hard, thermosetting, cross-linked polymer. The most commonly used epoxy resins are derived from diphenylolpropane (bisphenol A) and epichlorohydrin. Resins produced from these reactants are liquids that are too viscous to penetrate stone deeply. Therefore, they are diluted with organic solvents. These epoxy resins are often cured using an amine curing agent. Their cure time can be adjusted by selecting a slowly or rapidly reacting curing agent and by controlling the curing temperature. The resulting cross-linked polymers have excellent adhesion to stone and concrete and excellent chemical resistance. Lee and Neville, and Gauri, are recommended sources for information on the chemistry, curing, and applications of epoxies.[73,75]

Gauri developed a way to achieve deep penetration with viscous epoxy resins and at the same time to avoid the formation of a sharp interface between the consolidated and untreated stone.[75,76,100] Specimens were soaked in acetone, then in a dilute solution of epoxy resin in acetone, then in increasingly concentrated solutions. This method is feasible for tombstones and statues, but probably would be too time-consuming and expensive for large structures.

Less viscous epoxy resins are available, including diepoxybutane diglycidyl ether and butanediol diglycidyl ether.[50] Munnikendam cured butanediol diglycidyl ether with alicyclic polyamines such as menthane diamine. However, the viscosity was still too high, and he diluted the mixture with tetraethoxysilane and tetramethoxysilane. A reaction involving the epoxy resin, curing agent, and solvent took place to produce a tough, glassy material. A white efflorescence also developed from a reaction between the polyamine and carbon dioxide to form aminecarbonates.[59,77] Formation of the aminecarbonates can be avoided by preventing carbon dioxide from coming in contact with the solution before the desired reaction is complete. Gauri observed that when low-viscosity aliphatic epoxy resins were applied to calcareous stones, the rates of the reactions between the stones and carbon dioxide and sulfur dioxide were faster than the rates with untreated stones.[68,78] He suggested that the increased reactivity could be caused by absorption of the gases by the epoxy polymer or by the polymer acting as a semipermeable film to the gases. In contrast epoxy polymers based on bisphenol A were found to protect the stone from both carbon dioxide and sulfur dioxide.

The use of epoxies has been suggested for consolidating lime-stone,[1,68,69] marble,[75-80] and sandstone,[56,59] as well as for readhering large stone fragments to mass stone.[60] Hempel and Moncrieff found that certain epoxies could encapsulate salts in marble, thereby preventing them from recrystallizing.[80] A large restoration project using epoxies for masonry consolidation is that at the Santa Maria Maggiore Church in Venice.[81]

Like poly(methyl methacrylate), epoxies have produced brittle epoxy-impregnated concretes with high mechanical properties.[62,82,83] The long-term effect of incorporating a brittle material in stone is not known, but such a material could render a structure more vulnerable to seismic shock, vibrations, and effects of thermal expansion.

Many types of epoxies have a tendency to chalk (i.e., to form a white powdery surface) when exposed to sunlight.[73] Therefore, epoxy should be removed from the surface of a treated stone before it cures.

Other Synthetic Organic Polymers

Other synthetic organic polymers studied as possible stone consolidants include polyester,[67,84] polyurethane,[55] and nylon.[85] Polyester has been shown to decrease the porosity of stone substantially[84] and, therefore, may form an impervious layer that prevents the passage of entrapped moisture or salts.[67] Manaresi and Steen observed that polyurethanes were poor cementing agents.[56,86] Steen also found that a polyurethane film gradually became brittle when exposed to sunlight.[87] Similarly, DeWitte found that nylon can produce a brittle film on the surface of stone.[85]

Waxes

Waxes have been applied to stone for more than 2,000 years. Vitruvius described the impregnation of stone with wax in the first century B.C.[88] A wax dissolved in turpentine was one of several materials applied to the decaying stone of Westminster Abbey between 1857 and 1859.[89]

Cleopatra's Needle in London was first treated with wax in 1879 and has been treated several times since.[90] Kessler found that paraffin waxes were effective in increasing the water repellency of stone.[91] Waxes have also been found to be effective consolidants.[40,50,70,92] For example, a paraffin wax increased the tensile strength of a porous stone from 1.06 MN/m² (153 psi) to 4.12 MN/m² (594 psi), while triethoxy-methylsilane only increased it to 1.88 MN/m² (271 psi).[40,92] In addition, paraffin waxes are among the most durable stone conservation materials[7,70] and can immobilze soluble salts.[50]

Waxes have been applied to stone in solution in organic solvents,[7,9,90] by immersing a stone object in molten wax,[50] and by applying molten wax to preheated stone.[93] If deep penetration is not achieved, a non-porous surface layer may be formed, causing the eventual spalling of the treated surface.[9]

Major problems encountered in using waxes to conserve stone include their tendency to soften at high ambient temperatures[11] and to entrap dust and grime.[50,70,102] Wax applied to Cleopatra's Needle has gradually converted to a tarry substance which cannot be removed by ordinary washing. A mixture of carbon tetrachloride, benzene, and detergent was needed in 1947 to clean the Needle.[90]

COMMENTS ON STONE CONSOLIDANTS

Although stone consolidants have been used for more than a century, their selection is still based largely on empirical considerations. If a consolidant appears to give acceptable results with one type of stone, it is often applied to other types of stone without properly determining if it is compatible with them. Some of the factors affecting the performances of consolidants are known, such as depth of penetration and moisture transfer through consolidated stone. However, insufficient consideration has been given to equally important factors such as their consolidating abilities and the compatibility of their thermal expansion properties with those of stone. Finally, the long-term performances of consolidated stones in historic structures are rarely documented.

These considerations point to the inadequacy of the present state of stone consolidation and conservation technology. For example, stone consolidants should be selected on the basis of an understanding of the deterioration processes of stone and treated stone, of the factors affecting the performances of consolidants, and of the compatibility of consolidants with specific stones. Currently, such information often is not available. Further, standard test methods and performance criteria should be developed as a basis for selecting promising consolidants. Documentation of the performances of stone consolidants should be an integral part of each preservation or restoration program. Documenting unsuccessful consolidation work is just as important as documenting successful work in that it enables other stone conservators to reject ineffective materials and methods.

This review clearly indicates that a perfect stone consolidant has not been developed and that many of the proposed treatments can harm stone. Therefore, the general use of stone consolidants is open to question. In fact the British Commonwealth War Graves Commission, which is responsible for more than 1 million headstones in Europe, has concluded that no consolidant should be applied to headstones.[102] This commission has more than 50 years of experience with the chemical treatment of stone. There are cases, however, in which the use of stone consolidants can be beneficial. The work by Hempel and Moncrieff has shown that decaying stone statues can be preserved by deep impregnation with certain stone consolidants.[48,49,72,79,80] Statues and smaller objects can be removed to laboratories, thoroughly cleaned, freed from soluble salts, and treated on all sides with a consolidant, but such processes are not possible with masssive stone structures. The risks involved in treating massive structures, therefore, are greater. Consolidants might be used on structures of little historical or intrinsic value and in other cases where the benefits outweigh the risks involved.[102] For example, consolidants could be applied to deteriorated stone to delay the need to replace it with new stone. Any permanent consolidation effort involving important historic stone structures, however, should be carefully planned and carried out to minimize the risks. This includes making certain that moisture and soluble salts are not trapped behind the layer of treated stone. In addition, the compatibility of a consolidant with a specific stone should be determined with separate test specimens rather than by using an important historic structure as an experiment.

There is an obvious need for caution, even in the use of materials that have shown promise in accelerated laboratory tests. While accelerated tests designed in accord with ASTM Standard E 632 should be useful in the evaluation of stone consolidants, there will always be assumptions to be made about factors affecting performance. These assumptions will leave a measure of uncertainty about the reliability of predictions based on the test results, but the tests will minimize the risks in selecting a stone consolidant.

SUMMARY AND CONCLUSIONS

The main function of a stone consolidant is to reestablish the integrity of deteriorated stone by restoring intergranular bonds. In addition to consolidation, a stone consolidant should meet performance requirements concerning depth of penetration, compatibility with stone, effect on permeability and moisture transfer, effect on appearance, and durability. These may be termed "primary performance requirements" because they are applicable to all stone consolidants regardless of the specific use. Secondary performance requirements may sometimes have to be imposed because of specific problems encountered with certain structures. An example would be to require a consolidant to immobilize soluble salts in a stone.

In the selection of a consolidant many factors must be considered. These include the type of stone to be consolidated, the processes responsible for the deterioration of stone, the degree of deterioration, the environment, the amount of stone to be consolidated, and the importance of the structure. A universal consolidant does not exist because many of these factors will vary from structure to structure. Therefore, the preservation of each stone structure should be considered a unique problem.

Few cases of long-term success with consolidating stone structures were disclosed in this review. Some apparent success has been achieved in consolidating small stone objects, such as statues, which can be treated in a laboratory. Consolidants should be used on historic stone building or structure only after a careful appraisal has been made of the risks involved, the benefits to be realized, and the probability of success. ASTM Standard E 632 is a useful guide to considerations that should govern the development and use of accelerated tests for evaluating stone consolidants and building materials in general.

REFERENCES

1. E.M. Winkler, *Stone: Properties, Durability in Man's Environment*, 2nd edition (Springer-Verlag, New York, 1975).

2. J.R. Clifton, *Stone Consolidating Materials: A Status Report*, NBS Technical Note 1118 (National Bureau of Standards, Washington, D.C., 1980).

3. ASTM Standard E 632, *Recommended Practice for Development of Short-Term Accelerated Tests for Prediction of Service Life of Building Materials and Components* (American Society for Testing and Materials, Philadelphia, 1980).

4. G. Frohnsdorff and L.W. Masters, The Meaning of Durability and Durability Prediction, pp. 17–35 in *Durability of Building Materials and Components*, ASTM STP 691, P.J. Sereda and G.G. Litvan, eds., (American Society of Testing and Materials, Philadelphia, 1980).

5. A.M. Schwartz, Capillarity: Theory and Practice, pp. 2–13 in *Chemistry and Physics of Interfaces, II*, D.E. Gushee, ed., (American Chemical Society, Washington, D.C., 1971).

6. W.A. Zisman, ed., *Contact Angle, Wettability and Adhesion: Advances in Chemistry Series No. 43* (American Chemical Society, Washington, D.C., 1964).

7. A.R. Warnes, *Building Stones: Their Properties, Decay, and Preservation* (Ernest Benn, London, 1926).

8. I.E. Marsh, *Stone Decay and Its Prevention* (Basil Blackwell, Oxford, 1926).

9. A.P. Laurie and C. Ranken, The Preservation of Decaying Stone, *Journal of the Society of the Chemical Industry*, 37, pp. 137T–147T (1918).

10. G. Torraca, Brick, Adobe, Stone and Architectural Ceramics: Deterioration Processes and Conservation Practices, pp. 143–165, in reference 14.

11. N. Heaton, The Preservation of Stone, *Journal of the Royal Society of Arts*, 70, pp. 129–139 (1921).

12. J. Lehmann, Damage by Accumulation of Soluble Salts in Stonework, pp. 35–46, in reference 94.

13. D.S. Knopman, Conservation of Stone Artworks: Barely a Role for Science, *Science*, 190, pp. 1187–1188 (1975).

14. *Preservation and Conservation: Principles and Practices* (The Preservation Press, Washington, D.C., 1976).

15. A.P. Laurie, *Building Materials* (Oliver and Boyd, Edinburgh, 1922).

16. J.W. Mellor, *A Comprehensive Treatise on Inorganic and Theoretical Chemistry, Vol. VI: Silicates* (Longmans, Green and Co., London, 1925).

17. *Encyclopedia of Chemical Reactions*, Vol. VI (Reinhold, New York, 1956).

18. L. Kessler, A Process for Hardening Soft Limestone by Means of the Fluosilicates of Insoluble Oxides, *Comptes Rendus*, 96, pp. 1317–1319 (1883).

19. F.S. Barff, Stone, Artificial Stone, Preserving Stone, Colouring, British Patent 2608 (1860); from reference 97.

20. The Artificial Hardening of Soft Stone, *Stone*, 34, pp. 365–366 (1913).

21. J. Riederer, Further Progress in German Stone Conservation, pp. 369–385, in reference 60.

22. R. Wihr and G. Steenken, On the Preservation of Monuments and Works of Art with Silicates, pp. 71–75, in reference 94.

23. T. Stambolov, Conservation of Stone, pp. 119–124, in reference 94.

24. B.G. Shore, *Stones of Britain* (Leonard Itill Books, London, 1957).

25. B. Penkala, The Influence of Surface Protecting Agents on the Technical Properties of Stone, *Ochrona Zabytrow*, 17 (No. 1), pp. 37–43 (1964).

26. A.R. Powys, *Repair of Ancient Buildings* (J.M. Dent, London, 1929).

27. W. Linke, *Solubilities of Inorganic and Metal Organic Compounds, Vol. I* (Van Nostrand, Princeton, 1958).

28. H.G. Lloyd, Hardening Stone and Earth, British Patent 441,568 (1934).

29. V. Mankowsky, The Weathering of Our Large Monuments, *Die Denkmalpflege*, 12 (No. 1), pp. 51–54 (1910); from reference 97.

30. Breathing New Life into the Statues of Wells, *New Scientist*, pp. 754–756, (December 1977).

31. A.H. Church, Stone, Preserving and Colouring; Cements, British Patent 220 (1862); from reference 97.

32. A.H. Church, Treatment of Decayed Stone-Work in the Chapter House, Westminster Abbey, *Journal of the Society of Chemical Industry*, 23, p. 824 (1904).

33. A.H. Church, Conservation of Historic Buildings and Frescoes, *Proceedings of the Meetings of the Members, Royal Institution of Great Britain*, 18, pp. 597–608 (1907).

34. S.Z. Lewin, The Conservation of Limestone Objects and Structures, in *Study of Weathering of Stone, Vol. 1* (International Council of Monuments and Sites, Paris, 1968).

35. E.V. Sayre, Direct Deposition of Barium Sulfate from Homogeneous Solution Within Porous Stone, pp. 115–118, in reference 94.

36. G.G. Scott, Process as Applied to Rapidly-Decayed Stone in Westminster Abbey, *The Builder* (London), 19, p. 105 (1861); from reference 97.

37. S.Z. Lewin and N.S. Baer, Rationale of the Barium Hydroxide-Urea Treatment of Decayed Stone, *Studies in Conservation*, 19, pp. 24–35 (1974).

38. J.M. Garrido, The Portal of the Monastery of Santa Maria de Ripoll, *Monumentum*, 1, pp. 79–98 (1967).

39. G. Zava, B. Badan, and L. Marchesini, Structural Regeneration by Induced Mineralization of the Stone of Eraclea (Agrigento) Theatre, pp. 387–399, in reference 95.

40. L. Arnold, D.B. Honeyborne, and C.A. Price, Conservation of Natural Stone, *Chemistry and Industry*, pp. 345–347 (April 1976).

41. R.A. Munnikendam, Acrylic Monomer Systems for Stone Impregnation, pp. 15–18, in reference 94.

42. C.A. Price, Research on Natural Stone at the Building Research Establishment, *Natural Stone Directory* (1977).

43. Bosch, Use of Silicones in Conservation of Monuments, pp. 21–26, in *First International Symposium on the Deterioration of Building Stones* (1972).

44. H. Weber, Stone Renovation and Consolidation Using Silicones and Silicic Esters, pp. 375–385, in reference 95.

45. Preserving Building Stone, *BRE News*, 42, pp. 10–11 (Winter, 1977).

46. J. Taralon, C. Jaton, and G. Orial, Etat des Recherches Effectuees en France sur les Hydrofuges, *Studies in Conservation*, 20, pp. 455–476 (1975).

47. J. Riederer, Die Erhaltung Aegyptischer Baundenkmaeler, *Maltechnik Restauro*, 74 (No. 1), pp. 43–52 (1974).

48. A. Moncrieff, The Treatment of Deteriorating Stone with Silicone Resins: Interim Report, *Studies in Conservation*, Vol. 21, pp. 179–191 (1976).

49. K. Hempel and A. Moncrieff, Report on Work Since Last Meeting in Bologna, October 1971, pp. 319–339, in reference 95.

50. C.A. Price, Stone Decay and Preservation, *Chemistry in Britain*, 11 (No. 9), pp. 350–353 (1975).

51. A.P. Laurie, Preservation of Stone, U.S. Patent 1,607,762 (1926).

52. H.D. Cogan and C.A. Setterstrom, Ethyl Silicates, *Industrial Engineering Chemistry*, 39, pp. 1364–1368 (1947).

53. H.D. Cogan and C.A. Setterstrom, *Chemical Engineering News*, 24, pp. 2499–2501 (1946).

54. H. Marschner, Application of Salt Crystallization Test to Impregnated Stones, paper 3.4, in reference 98.

55. R. Snethlage and D.D. Klemm, Scanning Electron Microscope Investigations on Impregnated Sandstones, paper 5.7, in reference 98.

56. R. Rossi-Manaresi, Treatments for Sandstone Consolidation, pp. 547–571, in reference 95.

57. M.B. Caroe, Wells Cathedral, The West Front Conservation Programme: Interim Report on Aims and Techniques (June, 1977).

58. R.A. Munnikendam, Preliminary Notes on the Consolidation of Porous Building Stones by Impregnation with Monomers, *Studies in Conservation*, 12 (No. 4), pp. 158–162 (1967).

59. R.A. Munnikendam, A New System for the Consolidation of Fragile Stone, *Studies in Conservation*, 18, pp. 95–97 (1973).

60. G. Torraca, Treatment of Stone in Monuments. A Review of Principles and Processes, pp. 297–315, in *The Conservation of Stone, I: Proceedings of the International Symposium, Bologna, June 1975* (Centro per la Conservazione delle Sculture all'Aperto, Bologna, Italy, 1976).

61. L.E. Kukacka, A. Auskern, P. Colombo, J. Fontana, and M. Steinberg, *Introduc-*

tion to Concrete-Polymer Materials, Brookhaven National Laboratory. Available from National Technical Information Service, No. PB 241–691.

62. J. Clifton and G. Frohnsdorff, Polymer-Impregnated Concretes, pp. 174–196, in *Cements Research Progress 1975* (American Ceramic Society, Columbus, Ohio, 1976).

63. M. Steinberg et al., *Concrete-Polymer Materials, First Topical Report*, Brookhaven National Laboratory Report No. BNL–50134 (1968).

64. L.E. Kukacka et al., *Concrete-Polymer Materials, Fifth Topical Report*, Brookhaven National Laboratory Report No. BNL–50390 (1973).

65. E. Dahl-Jorgensen and W.F. Chen, *Stress-Strain Properties of Polymer Modified Concrete*, Fritz Engineering Laboratory Report No. FEL 390.1 (Lehigh University, Pennsylvania, 1973).

66. R.B. Seymour, *Introduction to Polymer Chemistry* (McGraw-Hill, New York, 1971).

67. H.A. LaFleur, *The Conservation of Stone. A Report on the Practical Aspects of a UNESCO Course, October and November 1976, Venice, Italy (National Park Service, Washington, D.C., 1977)*.

68. K.L. Gauri, P. Tanjaruphan, M.A. Rao, and T. Lipscomb, Reactivity of Treated and Untreated Marble in Carbon Dioxide Atmospheres, *Transactions of the Kentucky Academy of Science*, 38 (No. 1–2), pp. 38–44 (1977).

69. K.L. Gauri and M.V.A. Rao, Certain Epoxies, Fluorocarbon-Acrylics and Silicones as Stone Preservatives, pp. 73–80, in reference 99.

70. K.L. Gauri, J.A. Gwinn, and R.K. Popli, Performance Criteria for Stone Treatment, pp. 143–152, in reference 96.

71. Usefulness of Some Vinyl Polymers in the Conservation of Monuments, *Ochrona Zabytkow*, 14 (No. 3–4), pp. 81–92 (1961); from reference 97.

72. K.F.B. Hempel, Notes on the Conservation of Sculpture, Stone, Marble, and Terra-cotta, *Studies in Conservation*, 13, pp. 34–44 (1968).

73. H. Lee and K. Neville, *Handbook of Epoxy Resins* (McGraw-Hill, New York, 1967).

74. W.G. Potter, *Uses of Epoxy Resins* (Chemical Publishing Company, New York, 1975).

75. K.L. Gauri, Cleaning and Impregnation of Marble, in reference 100.

76. K.L. Gauri, Improved Impregnation Technique for the Preservation of Stone Statuary, *Nature*, 228 (No. 5274), p. 882 (1970).

77. G. Marinelli, Use of an Epoxy Aliphatic Resin in the Consolidation of Porous Building Materials Having Poor Mechnical Properties, pp. 573–591, in reference 95.

78. K.L. Gauri, Efficiency of Epoxy Resins as Stone Preservatives, *Studies in Conservation*, 19, pp. 100–101 (1974).

79. A. Moncrieff, Work on the Degeneration of Sculptural Stone, pp. 103–114, in reference 94.

80. K. Hempel and A. Moncrieff, Summary of Work on Marble Conservation at the Victoria and Albert Museum Conservation Department up to August 1971, in reference 100.

81. Epoxy Resin Saves Venice Church, *Corrosion Prevention and Control*, pp. 12–13 (October, 1974).

82. D.A. Whiting, P.R. Blankenhorn, and D.E. Kline, Effect of Hydration on the Mechanical Properties of Epoxy Impregnated Concrete, *Cement and Concrete Research*, 4 (No. 3), pp. 467–476 (1974).

83. D.A. Whiting, P.R. Blankenhorn, and D.E. Kline, Mechanical Properties of Epoxy Modified Concrete, *Journal of Testing and Evaluation*, 2 (No. 1), pp. 44–49 (1974).

84. M. Kranz, l'Evaluation de l'Etat de Conservation de la Pierre et de l'Efficacité des Traitements, pp. 443–453, in reference 95.

85. E. DeWitte, Soluble Nylon as Consolidation Agent for Stone, *Studies in Conservation*, 20 (No. 1), pp. 33–34 (1975).

86. C.R. Steen, Some Recent Experiments in Stabilizing Adobe and Stone, pp. 59–64, in reference 94.

87. C. Steen, An Archaeologist's Summary of Adobe, *El Palacio*, 77 (No. 4), pp. 29–38 (1971).

88. Vitruvius, *De Architectura*, I.X.; from reference 50.

89. G.G. Scott, Process as Applied to Rapidly-Decayed Stones in Westminster Abbey, *The Builder*, 19 (No. 941), pp. 105 (1861); from reference 97.

90. S.G. Burgess and R.J. Schaffer, Cleopatra's Needle, *Chemistry and Industry*, pp. 1026–1029 (1952).

91. D.W. Kessler, Exposure Tests on Waterproofing Colourless Materials, Technological Papers of the Bureau of Standards, No. 248, (1924–25).

92. L. Arnold and C.A. Price, The Laboratory Assessment of Stone, pp. 695–704, in reference 95.

93. The Preservation of Ruins, *Literary Digest*, pp. 94–95 (July, 1910).

94. *Conservation of New Stone and Wooden Objects, New York Conference, June 1970* (The International Institute for Conservation of Historic and Artistic Works, London).

95. *The Conservation of Stone, I, Proceedings of the International Symposium, Bologna, June 1975* (Centro per la Conservazione delle Sculture all'Aperto, Bologna, Italy, 1976).

96. Y. Efes and S. Luckat, Relations Between Corrosion of Sandstones and Uptake Rates of Air Pollutants at the Cologne Cathedral, pp. 193–200, *International Symposium on the Deterioration of Building Materials* (Athens, Greece, 1976).

97. S.Z. Lewin, The Preservation of Natural Stone, 1839–1965, An Annotated Bibliography, *Art and Archaeology Technical Abstracts*, 6 (No. 1), pp. 185–277 (1966).

98. Deterioration and Protection of Stone Monuments, International Symposium (Paris, June 1978).

99. Decay and Preservation of Stone, *Engineering Geology Case Histories No. 11*, E.M. Winkler, ed. (The Geological Society of America, Boulder, 1978).

100. *The Treatment of Stone, Proceedings of the Meeting of the Joint Committee for the Conservation of Stone, Bologna, October, 1971* (Centro per la Conservazione delle Sculture all'Aperto, Bologna, Italy, 1972).

101. *Stone Preservatives*, Digest First Series No. 128, Building Research Station (Garston, England, 1963).

102. W.H. Dukes, Conservation of Stone: Chemical Treatments, *The Architects' Journal Information Library*, pp. 433–438 (August 23, 1972).

The Suitability of
Polymer Composites as
Protective Materials

ANTHONY T. DIBENEDETTO

One method of minimizing the erosion of stone surfaces is to deposit into the surface pores organic monomers or prepolymers, which are then polymerized into a protective layer. This forms a stone–polymer composite near the surface that is meant to protect the stone from further erosion by water and reactive gases. The effectiveness of such a composite as a protective material depends on the mechanical properties of the composite, the stability of the interface between the stone and the polymer, and the permeation characteristics of the polymeric coating. This paper discusses the properties of composite systems relevant to the preservation of stone.

There are no materials of construction that are totally immune to environmental degradation. Even stone, undoubtedly the most durable of all traditional materials, is subject to physical and chemical erosion. The rate of degradation depends on both the type of stone and the nature of the environment. The realization that many historic structures are slowly deteriorating under centuries of exposure to atmospheric conditions, a deterioration often accelerated by the gaseous and particulate pollutants so common in today's urban environment, has led to the development of many novel techniques of preservation. Among those techniques is the application of synthetic organic mono-

---

Anthony T. DiBenedetto *is Professor of Chemical Engineering, University of Connecticut, Storrs.*

mers and polymers for the purpose of repairing, consolidating, and protecting the stone structures.

Many properties of polymers and polymer composites make them attractive for these purposes. For example, patching materials may be developed using polyester or epoxy resins filled with both stone particles and pigments, forming composite materials of strength, durability, and compatibility with the stone substrate. Consolidation and surface protection can be attained by impregnating the stone with low-viscosity monomers and subsequently polymerizing the monomers in the surface pores. This process forms coatings that resist water and gaseous-vapor penetration, in addition to strengthening the material.

The suitability of such treatment is strongly dependent on the properties of the composite materials formed. A few of the characteristics that determine whether the treatment will be helpful or harmful are the solids content of the composite, the adhesion between the resin matrix and the filler, the adhesion between the polymeric composite and the stone substrate, the matching of physical properties between the polymer composite and the stone, and especially the choice of resin matrix.

POLYMER COMPOSITES AS PATCHING MATERIALS

A low-viscosity polyester or epoxy prepolymer can be filled with crushed and graded stone particles and a pigment to produce an adhesive patching compound. Since it is essential to bond the resin to both the filler particles and the stone substrate, it is usually necessary to incorporate a bonding agent into the compound formulation. There are a number of such treatments, the most popular for polymer/silicate-type composites being the addition of a few percent of an organosilane. An organosilane coupling agent is a compound of the general form R-Si-$(X)_3$, where R is a resinophilic group and the X's are organic groups capable of interacting with silanols. Some examples of commercial organosilane coupling agents are listed in Table 1. For example, the three methoxy groups (OCH_3) of γ-glycidoxy-propyltrimethoxysilane (the seventh compound in Table 1) are capable of reacting with the hydroxyl groups of a silicate surface to form a chemically bonded layer of epoxy groups (CH_2CH-) on the inorganic surface (see Figure 1).

The other methoxy groups are also capable of reaction with the silicate or can interpolymerize in the presence of water to form a protective organosilane polymeric coating. The pendant epoxy group can react with an epoxy resin, so that the matrix is chemically bounded to the silicate substrate.

TABLE 1 Commercial Silane Coupling Agents

Name	Formula	Application
Vinyltriethoxysilane	$CH_2=CHSi(OC_2H_5)_3$	Unsaturated polymers
Vinyl-tris (β-methoxyethoxy)-silane	$CH_2=CHSi(OCH_2CH_2OCH_3)_3$	Unsaturated polymers
Vinyltriacetoxysilane	$CH_2=CHSi(OOCCH_3)_3$	Unsaturated polymers
γ-Methacryloxypropyltrimethoxysilane	$CH_2=C(CH_3)COO(CH_2)_3Si(OCH_3)_3$	Unsaturated polymers
γ-Aminopropyltriethoxysilane	$H_2NCH_2CH_2CH_2Si(OC_2H_5)_3$	Epoxies, phenolics, nylon
γ-[β-aminoethyl]aminopropyltrimethoxysilane	$H_2NCH_2CH_2NH(CH_2)_3Si(OCH_3)_3$	Epoxies, phenolics, nylon
γ-Glycidoxypropyltrimethoxysilane	$CH_2CHCH_2O(CH_2)_3Si(OCH_3)_3$	Almost all resins
γ-Mercaptopropyltrimethoxysilane	$HSCH_2CH_2CH_2Si(OCH_3)_3$	Almost all resins
β-[3,4-epoxycyclohexyl]ethyltrimethoxysilane	—$CH_2CH_2Si(OCH_3)_3$	Epoxies
γ-Chloropropyltrimethoxysilane	$ClCH_2CH_2CH_2Si(OMe)_3$	Epoxies

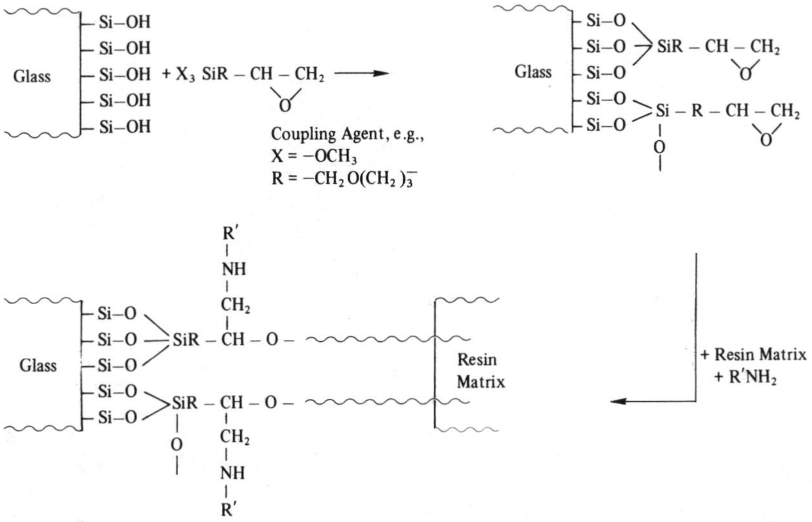

FIGURE 1 Coupling reaction at a glass/silane/epoxy interface.

Proper choice of a coupling agent is essential to the development of maximum physical properties and, perhaps more importantly, to long-term stability in the presence of moisture or soluble gases, such as sulfur dioxide. In the absence of good adhesion between the phases, thermal stresses will lead to debonding of the imbedded particles and a microcavitation of the materials. Water and gases can then collect at the interfacial void spaces, causing an accelerated degradation of the inorganic phase. This is illustrated schematically in Figure 2.

Even with the proper choice of components, the wettability of a surface must be considered. When a fluid polymer is placed on a silicate surface, it will form a contact angle, as shown schematically in Figure 3. A large contact angle, θ, represents poor wetting, while a contact angle of zero represents spontaneous wetting. Wetting is favored when the substrate is free of contamination, when the polymer has an affinity for the substrate, and when the surface tension of the polymer is low. Surface roughness affects the wetting characteristics, since the fluid must move up and over asperities. Most important is the possibility of air being trapped under a spreading fluid, thereby creating many voids at the adhesive interface. The inevitable thermal cycling of the material could then more easily lead to adhesive failure at the joint interface.

When patching a porous surface, a strong mechanical bond can be

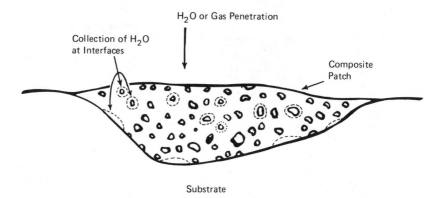

FIGURE 2 With poor adhesion between phases in a silicate-resin patching material, thermal stresses will lead to debonding of imbedded particles and microcavitation of materials. Water and gases can then collect at interfacial voids and cause accelerated degradation of inorganic phase.

developed by forcing the fluid into the capillary passages leading to the interior. First, the fluid must wet the capillary passages in order to displace the air in the pores (see Figure 4). Second, enough time must be allowed for penetration to occur. In a cylindrical, open pore of diameter, d, the depth of penetration is equal to:

$$\sqrt{\frac{\cos \theta \; \gamma_{LV} \, d \, t}{4\eta}}, \qquad (1)$$

where θ is the contact angle, γ_{LV} is the surface tension of the fluid, η is the viscosity of the fluid, and t is the time of penetration. Most patching compounds will be highly filled, so that the viscosity is extremely high, causing penetration to be very slow and in many cases negligible.

Even when the mechanical joint is perfectly made, one still has to accept the fact that there is a mismatch of physical properties between the patch and the substrate. Perhaps the most important is the difference in thermal expansion coefficients. The polymer composite will always have a higher thermal expansion coefficient, and thus normal thermal cycling will lead to internal stresses at all particle/polymer interfaces and at the composite/substrate boundaries. Under adverse conditions this could lead to microcracking of the patch material and even accelerated damage to the substrate. Thus, a great deal of care

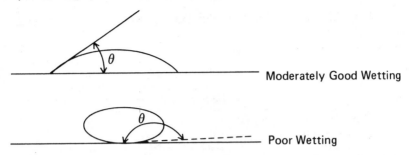

FIGURE 3 Contact angle is a measure of wettability. Fluid polymer forms contact angle (θ) at a surface. The smaller the contact angle, the better the wetting; a contact angle of zero represents spontaneous wetting. In this schematic, LV is the liquid–vapor interface, SL is the solid–liquid interface, and SV is the solid–vapor interface.

must be taken in choosing the proper patching compound for a given application.

A critical variable is the choice of polymer matrix. Polyester and epoxy are generic names for a multitude of different resin formulations with vast differences in both physical properties and environmental stability. Some polyesters, for example, have relatively low resistance to atmospheric humidity, while others are highly resistant.

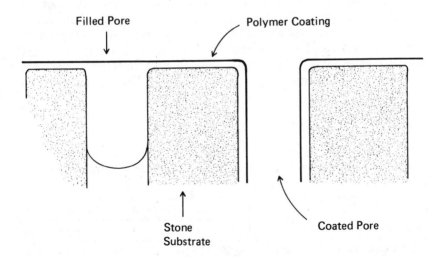

FIGURE 4 Bonding strengthened by forcing polymer into pores. Patching material can form strong mechanical bond with porous substrate when forced into capillary passages. Depth of penetration depends on contact angle, surface tension of fluid, viscosity of fluid, and time of penetration (see equation 1 and Figure 3).

TABLE 2 Illustrative Sorption and Transmission Rates of Water in Polymers at 40° C

Polymer	% H_2O 24-hour immersion 1/8-inch-thick sample	Water transmission rate 90–95% RH (g/m²/24 hours/mil)
Polyethylene (0.92 g/ml)	0.01	28
Polyethylene (0.96 g/ml)	0.01	4
Poly(vinyl chloride)	0.03	32
Poly(vinyl choride) (plasticized)	0.4	88
Poly(methyl methacrylate)	0.2–0.4	550
Poly(ethylene terephthalate) (Mylar)	0.03–2.5	30
Silicone rubber	0.1–0.15	—
Epoxy resin	0.01–0.5	—

The five varieties of polyester resins commercially available are (1) general purpose, (2) isophthalic polyester, (3) bisphenol-A-based polyester, (4) chlorine-bearing polyesters, and (5) vinyl ester resins.

The general purpose types, composed of phthalic anhydride, maleic anhydride, and propylene glycol, are the lowest-cost resins, but they generally offer the least corrosion resistance. The isophthalic types use a phthalic acid monomer and exhibit better resistance in both acidic environments and saltwater. The other types have higher temperature resistance and can be formulated for chemical resistance under very severe conditions. The bisphenol-A-based polyesters are commonly used and are highly resistant to a wide variety of harsh environmental conditions.

TABLE 3 Some Gas Permeation Values Through Polymer Films at 30° C

Film	$P \times 10^{10}$ (cc (STP)/mm/cm²/sec/cm Hg)			
	O^2	H_2S	CO_2	H_2O
Poly(vinylidene chloride) (Saran)	0.05	0.29	0.31	15–100
Polyester (Mylar A)	0.22	0.71	1.50	1300
Polychlorotrifluoroethylene (Kel-F)	5.60	—	12.50	2.9
Polyethylene (0.92 g/ml)	55.0	448.0	350.0	800
Natural rubber	230.0	1200.0	1330.0	—

The epoxy resins are formed by the reaction of epichlorohydrin with a hydroxyl-containing compound, such as bisphenol A, and then cured to a thermoset material with anhydride or amine curing agents. They have somewhat better thermal properties than the polyesters and are generally more resistant to corrosion, except in the presence of strong oxidizing agents. Once again, however, the specific properties are highly dependent on the monomer formulation. The epoxy chain can be aromatic-based (starting with bisphenol A) or cycloaliphatic or aliphatic (starting with glycerol). These prepolymers can be cured with long-chain or short-chain amines or a variety of acid or anhydride catalysts. Other commercially available formulations are simply too numerous to mention. It is possible to formulate epoxies with softening points ranging from 50° C to 200° C, with water solubilities ranging from less than 0.1 percent to more than 5 percent, with gas-transmission rates varying by orders of magnitude, and with mechanical properties ranging from high ductility and impact resistance to extreme brittleness.

Water and gas permeability depend on both the solubility and the diffusivity of the penetrant in the polymer material. Both of these properties can be varied over wide ranges by choosing the prepolymer components appropriately. Tables 2 and 3 give some idea of the range of gas and water vapor transmission rates possible with different types of polymer films. In any particular application these properties would have to be measured for the compounds being considered as treating agents. Epoxy formulations, for example, could be highly aromatic or highly aliphatic and have properties at either end of the spectrum illustrated by Tables 2 and 3.

CONSOLIDATION OF STONE USING MONOMERS AND PREPOLYMERS

An increasingly popular method of protecting and strengthening the surface of a stone structure is to impregnate the surface with an organic monomer or prepolymer and then to polymerize the material in the surface pores of the stone. If this is done properly, the monomer will coat deeply the internal surface pores, adhere to them, and then polymerize into a tough polymer film with low permeability to water and corrosive gases. Many polymers have been tried, including epoxies, silicones, fluorocarbons and poly(methyl methacrylate). The consolidated surface is a composite structure the properties of which depend on the constituent phases as well as on the adhesion between the two phases.

The first step of the impregnation process requires that the monomer penetrate the stone pores. As previously expressed by equation 1, the depth of penetration is a function of contact angle, surface tension, viscosity, pore size, and time. Very often, some of the properties desirable in the consolidated structure cause difficulties in the processing. For example, a silicone might be desirable because of its water resistance, while that very characteristic may mean it is relatively nonwetting (high contact angle). Thus, it may be difficult to make a silicone penetrate the surface pores, which would result in shallow impregnation, clogging of pores, and a surface film that could be subject to peeling. Some of the most satisfactory epoxy prepolymers also have very high viscosities, which decrease the depth of penetration. Under certain conditions it might be necessary to dilute the epoxy with either a reactive or nonreactive solvent to permit penetration. Upon curing, however, the presence of a solvent will result in greater shrinkage, leading to higher internal stresses and, perhaps, cracking of both the polymer coating and the stone. Thermoplastic materials, such as poly(methyl methacrylate), have a combination of good wetting characteristics, mechanical strength, and impermeability and can be applied in a low-viscosity monomeric form. Upon long-term exposure to wet, polluted air, and under the internal stresses created at the pore surfaces, however, there can be a tendency for the polymeric coating to craze, thus resulting in loss of protection.

The above-mentioned difficulties should not discourage the use of polymers in stone consolidation. It is possible to choose a resin with proper characteristics for both processing and long-term stability. The environmental stability, so strongly dependent on the permeability characteristics, is determined by choosing the proper resin for the known environmental conditions.

When choosing a material for consolidation, one should seek expert advice. Also, regardless of the degree of expertise, one should expect to obtain and analyze accelerated test data on the composite system. The improper choice of consolidation materials can lead to a bigger problem than was there originally. Our historic structures are too important for us to make hasty decisions on the means of saving them.

BIBLIOGRAPHY

Anonymous. *Chemistry and Physics of Interfaces*, Americal Chemical Society: Washington, D.C., 1965.

DePuy, G.W., L.E. Kukacka, A. Aushern, W.C. Cowan, P. Colombo, W.T. Lockman, A.J. Romano, W.G. Smoak, M. Steinberg, F.E. Causey. *Concrete-Polymer Materials, Fifth Topical Report*, BNL 50390 and USBR REC–ERC–73–12, 1973.

Gauri, K. Lal. The preservation of stone. *Scientific American*, June 1978.

Manson, J.A., and L.H. Sperling. Polymer Blends and Composites, pp. 335–371 in *Filled Porous Systems*. Plenum Press: New York, 1976.

Sternman, S., and J.G. Marsden. Bonding Organic Polymers to Glass by Silane Coupling Agents. In *Fundamental Aspects of Fiber Reinforced Plastic Composites*, R.T. Schwartz and H.S. Schwartz, eds. Interscience: New York, 1968.

Exposure Site and Weatherometer Evaluations of Synthetic Polymers

W. LINCOLN HAWKINS

The environmental stability of synthetic polymers is important in applying them as protective coatings or sealants for historic stone buildings and monuments. These materials are attractive as protectants because of their ability to form highly impervious films or to penetrate into ceramic materials and so function as sealants. Care must be exercised in selection of the synthetic polymer having the highest level of stability under conditions to be encountered, while at the same time not losing other important properties. The addition of stabilizers can do much to extend the useful life of synthetic polymers, and test procedures have been developed to predict the stability of these materials during outdoor exposure.

Synthetic polymers are often used as coatings to protect buildings and other structures from the damaging effects of weathering. The use of paints, many of which have a polymeric component, to protect wooden structures dates back to the time of the pharaohs. Acrylic finishes for wood or metal are more modern applications, as is impregnation of stone and concrete with synthetic polymers, which is often employed as a technique for sealing out moisture. Despite these applications, however, the vulnerability of polymer coatings to various elements of the environment is often overlooked.

W. Lincoln Hawkins *is Director of Research, Plastics Institute of America, Hoboken, New Jersey.*

Almost all synthetic polymers are organic in composition. The only important exceptions are the siloxanes, and although these polymers are basically compounds of silicon and oxygen, they do contain carbon groupings in their structures. Most organic compounds, and polymers in particular, react readily with oxygen under a variety of conditions. In the case of synthetic polymers, oxidation usually results in rapid degradation, and properties important to protective film applications are adversely affected. Moisture also degrades certain polymers by breaking molecular chains into smaller fragments.

The stability of synthetic polymers in the face of outdoor weathering varies widely. Degradation depends, in the first instance, on the presence of chemical groups in the polymer molecule that absorb ultraviolet radiation in the frequency range to which the polymer will be exposed. Polyethylene, which is related to paraffin wax, should be transparent to ultraviolet radiation. However, this polymer has poor stability against weathering, apparently the result of oxygen-containing groups present as imperfections in the polymer molecules. Polymers with ring structures—polystyrene is an important example—absorb strongly in the ultraviolet range and yellow rapidly during outdoor exposure. Another familiar polymer, poly(vinyl chloride), absorbs ultraviolet radiation as a result of unsaturated groups in the polymer molecules and discolors extensively when exposed out-of-doors.

Poly(methyl methacrylate), marketed under the trade names of Lucite and Plexiglas, has very good resistance to ultraviolet radiation. As a result, it is used as a glazing material. Although polytetrafluoroethylene has very good resistance to weathering, it is difficult to fabricate into protective films. The siloxanes are among the most stable polymers in outdoor applications and also function well as moisture repellents.

FACTORS IN WEATHERING

It is important to understand the complex factors that are responsible for loss of stability and eventual failure of synthetic polymers during weathering. A variety of reactive chemical agents are present in the outdoor environment, and many of these contribute to degradation. Oxygen, the most pervasive of these reactants, attacks all polymers, although stability to oxygen varies considerably for individual polymers. In the form of ozone, oxygen may be an important catalyst for oxidative degradation, although this has not been clearly established.

Moisture is the second most damaging reactant in the atmosphere. Loss of stability by hydrolysis is very damaging to nylons, polyesters, and similar polymers made by condensation reactions. Polyethylene, polystyrene, and the polyacrylates (Lucite or Plexiglas) are resistant to attack by moisture.

Catalysts, which are present in the atmosphere at all geographic locations and particularly in industrial areas, can rapidly promote loss of stability. These catalysts include oxides of sulfur and nitrogen, ozone in abnormal concentrations, and organic peroxides from automobile exhausts. These catalysts may accelerate oxidation and, when acidic or basic, can cause the rapid hydrolysis of such polymers as nylons and polyesters. Thus, these polymers will degrade rapidly in contact with the "acid rain" that may occur in heavily industrialized areas. When such polymers are used as impregnants for stone structures, acidic (or basic) leachates from the stone or from concrete may cause them to degrade rapidly.

Adverse weathering may appear to be slow, but the rate increases rapidly as conditions are intensified. Heat from solar radiation is a primary energy source that promotes the loss of stability during outdoor exposure. Ultraviolet radiation also contributes to the degradation of most synthetic polymers.

The protective ozone layer in the upper atmosphere screens out ultraviolet radiation of frequencies below about 2900 Å. However, most polymers are degraded by ultraviolet radiation between 3000 and 3500 Å, and considerable energy within this frequency range reaches most areas of the earth's surface.

Thus, oxygen, moisture, many of the catalysts in the atmosphere, and solar heat and radiation constitute a very damaging combination of reactants. The adverse effects of weathering are the principal limitation on the use of polymers out-of-doors.

ANTIWEATHERING STABILIZERS

Considerable research has been directed toward developing stabilizers to offset adverse weathering effects. The chemical reactions responsible for photodegradation have been studied extensively, and as a result very effective stabilizers are now available. By prudent selection of stabilizers, the useful life of polymers can be extended considerably.

Thermal oxidation of polyethylene, polypropylene, and similar polymers occurs by a free-radical-initiated chain reaction. Once reaction is initiated at some site within the material, degradation continues by autocatalysis, in which a chain reaction results in further attack at hundreds of additional sites. Understanding of this mechanism has led to development of two types of thermal stabilizers or antioxidants.

The first of these inhibits free-radical initiation, and the second interrupts the chain reaction so that a few steps, rather than hundreds, result from an initiation reaction. For instance, unprotected polyethylene is completely degraded in about two years at temperatures slightly above ambient. When properly stabilized, however, this polymer resists degradation for 40–50 years at similar temperatures in accelerated tests. The life of polyethylene would be even greater under natural outdoor conditions because temperatures are usually much lower at night. Many of the newer stabilizers can be used in clear formulations; they do not impart color to the polymers. Effective protection can be realized with only a tenth of a percent of added stabilizer.

Protection against ultraviolet radiation is provided by pigments, which screen out the radiation, or by ultraviolet absorbers, which absorb the damaging radiation and then dissipate the absorbed energy in a manner not harmful to the polymer. Absorbed energy may dissipate through fluorescence, emission of visible light, or by chemical reactions of the stabilizer. Carbon black is by far the most effective of all available light screens. However, its use in protective films is limited to those applications in which black would be an acceptable finish. Ultraviolet absorbers, on the other hand, afford good protection for clear or light-colored films. Although not as effective as carbon black, several of the newer ultraviolet absorbers can extend the outdoor exposure life of some polymers by an order of magnitude.

When stability is lost and failure occurs during out-of-doors exposure, the extent to which thermal energy may have contributed to photooxidation must be considered. It is difficult to distinguish between concurrent reactions and the effects resulting from each. For this reason, combinations of an ultraviolet absorber and a thermal antioxidant are used when maximum protection against weathering is required. These combinations often result in a synergistic effect, giving better protection than would be anticipated from the effectiveness of the individual components.

ACCELERATED TESTS

Accelerated tests for stability against weathering fall into two categories, outdoor and indoor tests. Outdoor exposure is the most reliable test, but the time required to measure the stability of well-protected polymers is too long to be practical. Acceleration by a factor of about two can be realized through location of exposure sites in areas of intense solar radiation. In the United States, locations in the Southwest (the Sun Belt) and in Florida are preferred. The United Kingdom uses exposure sites in South Africa and similar areas. However, site location alone does not give sufficient acceleration. Conventional exposure sites employ an angle of 45° facing south for test samples. This fixed angle does not give the maximum exposure but has been adopted as a standard in order to make possible a reasonable level of cross-checking between different locations. A higher degree of acceleration is obtained by rotating test samples so as to follow the sun's path across the sky. These solar-tracking devices can reduce the test time by perhaps an additional factor of two. The most advanced of the solar-tracking devices uses a group of polished metal mirrors to accumulate solar radiation and concentrate the energy on test samples. The combined effects of solar tracking and accumulation of solar radiation have been

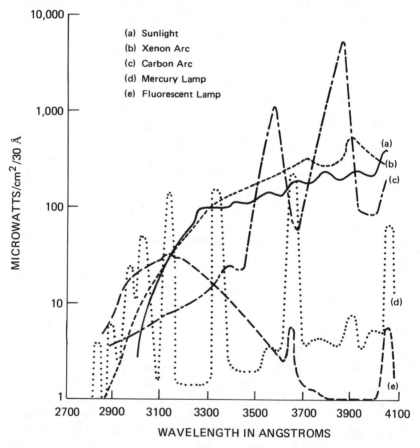

FIGURE 1   Comparative spectra of ultraviolet light sources. SOURCE: John Wiley & Sons.

claimed to give an acceleration factor as high as 11 over exposure at a fixed angle in a temperate zone. These methods may, however, provide an improper balance between thermal and radiation damage. It is difficult to maintain the temperature of test samples near what it would be under normal exposure conditions. And despite all the advances made in outdoor test procedures, exposure time is lost at night.

In an attempt to obtain 24-hour exposure, indoor tests have been developed that use an artificial ultraviolet source. Ultraviolet lamps and arcs have been used in indoor testing in devices known by the generic term "weatherometer." Weatherometers are further classified by the ultraviolet light source used. The comparative spectra of several common ultraviolet sources are compared with the solar spectrum in Figure 1. It is apparent from these data that the xenon arc corresponds most closely with the solar spectrum. Unless there is a close match

TABLE 1 Comparative Stability of Polymers During Outdoor Weathering (E—excellent, G—good, F—fair)

Polymer	Ultraviolet Radiation	Heat	Moisture
Polytetrafluoroethylene (Teflon)	E	E	E
Urea-formaldehyde resin (Melamine)	E	E	E
Poly(dimethoxy siloxane)	E	E	E
Poly(methyl methacrylate) (Lucite or Plexiglas)	E	G	G
Epoxy resins	G	G	G
Polyurethanes Ester type	F	F	F
Ether type	F	F	F
Poly(ethylene terephthalate) (Hytrel)	F	F	G
Polyethylene (Polythene or Alathon)	F	F	E
Polycarbonates (Lexan)	F	F	G
Nylons (Zytel)	F	F	F
Poly(vinyl chloride) (Exon, Geon, Tygon)	F	F	G

NOTE: These relative stabilities are for unprotected polymers. Selected trade names appear under the technical names and are in parentheses. Some of the polymers listed are generic names, and there may be significant differences between members in a family.

between the solar spectrum and that of the artificial light source, it is not possible to make an accurate prediction of weatherability.

In Table 1, polymers likely to be used as protective coatings are grouped according to their relative stabilities during outdoor exposure. The data shown are for unprotected polymers. Relative ratings could be altered somewhat for stabilized polymer compositions.

CONCLUSION

This paper is intended to alert potential users of synthetic polymers of the vulnerability of these potentially useful materials to degradation under ordinary outdoor exposure. Careful choice of polymers to be used is most important. Much more can be gained by incorporating an efficient stabilizer system into the base polymer. The final caution to be raised concerns a proper test procedure for a selected polymer composition, one that will give a reasonable prediction of the time to failure of the protective coating.

BIBLIOGRAPHY

Grassie, N., ed. *Chemistry of High Polymer Degradation Processes*. Wiley-Interscience: New York, 1956.

Hawkins, W.L., ed. *Polymer Stabilization*. Wiley-Interscience: New York, 1972.

Kamal, M.K., ed. *Applied Polymer Symposium No. 4*. Wiley-Interscience: New York, 1967.

Lundberg, W.O., ed. *Autoxidation and Antioxidants*. Wiley-Interscience: New York, 1961.

Scott, G., ed. *Atmospheric Oxidation and Antioxidants*. Elsevier: New York, 1965.

The Evaluation of Stone Preservatives

CLIFFORD A. PRICE

This paper reviews approaches to the evaluation of stone preservatives and groups them into categories: evaluation without experiment, evaluation by natural weathering, tests intended primarily for use on untreated stone, tests intended specifically for treated stone, and performance criteria. The paper does not review individual techniques in detail. Instead, it examines the thinking that underlies the techniques and questions both the validity of some procedures and their relevance to practical conservation problems. It concludes by outlining a policy for the use of stone preservatives.

The relentless deterioration of ancient monuments and historic buildings is a matter of pressing concern throughout the world. Numerous conservation strategies have been proposed, often involving the application of a preservative. Preservatives typically are colorless liquids applied to the building fabric in an attempt to strengthen that fabric and to protect it against further decay. The search for reliable preservatives has extended over many years but has been largely fruitless: No treatments have been found that are widely applicable and that have been proved effective over periods of several decades.

Clifford A. Price *is Head of the Stone Conservation Section, Building Research Establishment, U.K. Department of the Environment, Watford, England.*

This paper has been written as part of the research program of the Building Research Establishment of the Department of the Environment, United Kingdom, and is published by permission of the Director (Crown Copyright 1980, Building Research Establishment, Department of the Environment).

The increasing involvement of scientists in conservation problems is giving rise to many new suggestions for preservatives. Some of these materials (organosilicon compounds, for example) have been available only in recent years, and there is no long-term experience of their effectiveness. It would be grossly irresponsible to apply any unproven material to masonry of high artistic or historical importance, and yet it is just such masonry that is in the most urgent need of treatment. Reliable procedures for the rapid evaluation of potential preservatives are therefore essential.

The development of evaluation procedures is by no means straight-forward. Historic buildings and monuments are constructed from a bewildering variety of materials and are exposed to a comparable variety of environmental conditions. Preservative treatments accordingly must take many forms to counteract a wide range of decay mechanisms. Just as there can be no "universal preservative," there can be no universal evaluation procedure.

A glance through the proceedings of conservation conferences will reveal the many techniques that have been adopted by researchers for evaluating preservatives. It is not the purpose of this paper to review in detail the techniques themselves. Instead, the paper sets out to examine the thinking that underlies the techniques and to question both the validity of some evaluation procedures and their relevance to practical conservation problems. For convenience, the paper groups current evaluation procedures into a number of different categories, which correspond roughly with increasing levels of sophistication. The paper is unashamedly a personal view, and the categorization is only one of many that would be possible. Inevitably, there are areas of overlap between categories, and there is no suggestion that any one individual or institution will operate deliberately within one category alone. On the contrary, most researchers will draw from most categories during their evaluation of any particular preservative.

The evaluation of novel materials and procedures is by no means exclusive to conservation scientists. It is hoped that this overview of the problem as it confronts conservation will stimulate discussion and invite comment from those with relevant experience in other fields.

Although the paper deals specifically with the evaluation of preservatives for stone, much of it is equally relevant to other porous building materials.

EVALUATION WITHOUT EXPERIMENT

The evaluation of preservatives without recourse to experiment represents perhaps the lowest level of sophistication. Nevertheless, it can play an important role. It is practiced, unconsciously maybe, when a preservative is rejected on grounds of past experience or of existing scientific knowledge. For example, poly(ethylene glycol) is sometimes used to consolidate stone carvings before indoor exhibition. It could reasonably be rejected without experiment for outdoor use since it is soluble in water. This type of evaluation—leading to rejection—is perfectly acceptable, even if somewhat trivial. On the other hand, evaluation without experiment is not acceptable when it argues for the acceptance of a preservative.

Although this sounds obvious, it is not always clearly seen. In England, for example, treatments based on lime are being used on several important limestone buildings, notably Wells and Exeter cathedrals. Some protagonists point out that the treatment is based on the very material from which the stone is derived; they argue that the treatment must be safe and effective since it puts back into the stone the same substance that it has lost through decay. (By the same token, they argue that organic impregnants must be harmful since they introduce materials foreign to the stone.) The argument has no logical validity, for the microstructure of lime-treated stone may be quite different from that of the original stone. Nevertheless, the argument has enormous emotional appeal and can easily sway the nonscientist.

This example is not intended to cast a slur on the treatment itself, which has so far proved extremely successful. It is the underlying argument tacitly adopted by some of the treatment's advocates that is under criticism here. The argument is reminiscent of the saying "I've never tried it, because I don't like it," except that here it is put in the converse: "Because I like it, I don't need to test it."

EVALUATION BY NATURAL WEATHERING

The evaluation of preservatives by exposing treated specimens to the weather is an obvious procedure. In the long term it is also the best, for no other procedure can carry 100 percent assurance of demonstrating the performance of preservatives under natural weathering.

Natural exposure trials fall into two categories. On the one hand, small blocks of stone may be exposed to the weather and brought periodically into the laboratory for washing and detailed examination. On the other hand, trials may involve the treatment of stonework that is an integral part of a building. Tests of the first type are surprisingly sensitive and may give a preliminary indication of a treatment's effectiveness after only a few months. On the roof of St. Paul's Cathedral in London, for example, exposure trials are in progress using limestone specimens approximately 100 mm × 80 mm × 60 mm. Untreated blocks lose approximately 1 g (0.1 percent) every 6 months, which is easily detected on a laboratory balance. The particular drawback of this type of test is that it may not accurately reflect the way that the treatment would behave in a building, where joints and structural movement will have an influence.

Preservative trials on buildings can range from the treatment of a few blocks to treatment of an entire facade. The condition of the experimental panel is carefully recorded before treatment, as is the condition of an adjacent panel that is to be left untreated. The comparative condition of the two panels thereafter provides an indication of the treatment's effectiveness.

The comparison is where difficulties may arise. Unless the performance of the two panels is so different that there is no possible scope for misinterpretation, a quantitative and objective method is required for monitoring the condition of the panels. Surprisingly little attention has been given to this area. Individual tests, such as rebound tests or measurement of surface hardness, have been used, but the literature has carried few accounts of systematic, quantitative monitoring. One of the reasons may be the frequent requirement that tests be nondestructive and also the necessity for panels to be large enough to permit statistically meaningful analysis. In practice, monitoring tends instead to be qualitative and somewhat subjective. It will be based, for example, on appraisals of surface hardness made by running the finger across the two panels and comparing the amounts of loose stone that it removes—the test is reasonably satisfactory at first but meaningless once all the loose stone has been removed from the control panel.

Despite the undeniable benefits of evaluation by natural exposure, there is a further drawback. A single trial can provide, at best, information on the behavior of preservatives on one type of stone in one particular environment. Extrapolation to other stones and environments is unreliable. To set up a thorough appraisal of a preservative on a range of stones under a variety of conditions of exposure is enormously time-consuming and costly. And when the trials have been established, significant results may not emerge for many years—by which time new preservatives will have been developed, and the whole sequence must start again.

To summarize, natural exposure has a central part to play in the evaluation of preservatives. Under no circumstances can it be omitted, for it is the only "true" test. Nevertheless, its shortcomings are such that there is an inevitable demand for accelerated laboratory tests. These are the subject of succeeding sections.

TESTS PRIMARILY FOR UNTREATED STONE

Many testing procedures are available that were developed originally for use on untreated stone. Some of these procedures measure properties that are directly relevant to the use of stone in buildings. For example, bending strength is relevant to a stone's suitability for use in lintels, and coefficients of expansion are needed to make allowance at the design.stage for movement. On the other hand, most test procedures are intended to provide guidance on a poorly defined and elusive property: durability. The procedures either simulate natural decay mechanisms (such as salt crystallization and freezing tests) or measure physical properties that may be expected to have a bearing on durability (such as pore structure).

It is understandable that durability tests for untreated stone should be considered when seeking tests for evaluating preservatives. Since it has been shown that coarse-pored stones are generally more durable than fine-pored stones, it is reasonable to expect that a treatment that decreases the proportion of fine pores should improve a stone's weathering resistance. Similarly, a treatment that improves a stone's performance in a crystallization test might be expected to increase resistance to salt crystallization in a building. However, matters are seldom so simple. Almost invariably the treatment will affect the test results in such a way as to make them highly questionable, if meaningful at all. As a case in point, porosimetry techniques frequently require knowledge of the contact angle between stone and invading fluid. Unless due allowance is made for the change in contact angle upon treatment, direct comparison of results is meaningless. The interpretation of results is further complicated if, as is often the case, the contact angle for the untreated stone lies on one side of 90° and the angle with the treated stone lies on the other. Similarly, water repellency in treated specimens can nullify the significance of the salt crystallization test.

It is unfortunate that a great many investigators have overlooked these difficulties when evaluating preservatives. Even when the problems have been recognized and ways of overcoming them proposed, the modified test procedures may still be unsatisfactory. Vacuum saturation, for instance, is sometimes proposed to overcome the water repellency of specimens in the crystallization test. However, this approach is unsatisfactory because the solution still will be unable to enter pores with an effective diameter below a certain value. It is essential that a very rigorous scrutiny be made of any test method intended primarily for use on untreated stone if the procedure is to be extrapolated to evaluate treated stone. In some cases the procedure will be equally meaningful on treated and untreated stone. In the majority, it will not.

TESTS SPECIFICALLY FOR TREATED STONE

Tests intended specifically for treated stone may be grouped into three categories: those aimed simply at characterizing the extent and nature of the treatment, those using accelerated aging chambers, and those aimed at assessing the extent to which treatments meet stated objectives. They will be considered in turn.

Characterization of Treatment

It is often desirable to record certain characteristics of a treatment even if they provide no direct information on the treatment's effectiveness. These characteristics may include the depth to which the treatment has penetrated the stone, the amount by which it has filled the pore space, the degree of polymerization (if relevant and practicable), and the effect that the treatment has had on the stone's appearance. Data like these are needed to build up a full picture of a treatment but, because they do not relate directly to effectiveness, they are not considered in detail here.

Accelerated Aging Chambers

Accelerated aging chambers frequently are used to evaluate the durability of organic construction materials like paints and plastics. They have been applied by a number of researchers to the evaluation of stone preservatives. The chambers typically are capable of simulating a variety of environmental conditions, either singly or in combination. The conditions may include wetting and drying, heating and cooling, exposure to ultraviolet and infrared radiation, and immersion in dilute acids and salt solutions. The results of tests in these chambers certainly are useful in indicating the relative resistance of different treatments to specific environments, although some of the problems of judging tests for untreated stone may also apply here.

The main drawback of accelerated aging chambers is that they cannot give 100 percent assurance that they accurately reflect the proper balance and interaction of natural weathering agents. Nor can there be any reliable indication of the "acceleration factor." In other words, at the end of a test, one is obliged to ask, "So what? Does the test really show what would happen to the specimens after *n* years of natural weathering? And how large is *n*?"

Accelerated aging chambers certainly must not be seen as all-sufficient. Instead, they should be seen as one element of the next category of tests—assessing how well treatments meet their objectives. In this context such chambers can be very useful, particularly when used to simulate just one set of environmental conditions.

Tests for Assessing How Well Treatments Meet Objectives

Instead of looking for an overall, nonspecific preservative action, one may more rationally assess the extent to which a treatment fulfills its stated objectives. One is then making a quantitative measurement of a clearly defined parameter, rather than appraising an ill-defined concept. Such an approach has the additional advantage that it can directly take into account differing decay mechanisms. Some examples will help to clarify the approach.

One of the functions of a stone preservative is to strengthen weathered stone, both to protect it against mechanical damage and to make it more resistant to decay mechanisms. One logical element of an evaluation procedure, therefore, must be the measurement of strength, both before and after the treatment is applied. Strength in tension is normally the most appropriate parameter, since the majority of decay mechanisms place the stone in tension.

Another property frequently required in a preservative is the ability to protect limestone from attack by acidic air pollutants. The capability of a preservative in this respect may be determined, for example, by exposing treated specimens to moist air containing sulfur dioxide. The amount of calcium sulfite or sulfate formed is an inverse measure of the treatment's effectiveness.

A final example concerns the ability of a preservative to prevent damage from salt crystallization. In principle, a suitable test would entail treating salt-contaminated specimens with the preservative and then subjecting the specimens to relative humidities that fluctuated above and below the equilibrium relative humidity for that particular salt.

This approach may also be used to assess the vulnerability of preservatives to specific risks. One such risk might be degradation of the preservative under ultraviolet light, a risk that may readily be evaluated in an accelerated aging chamber. Another example concerns the possibility of fracture at the interface between treated and untreated stone as a result of differential moisture and thermal movement. This risk can be investigated by treating specimens on one face only and then subjecting them to cycles of oven drying followed by rapid immersion in cold water. An associated investigation might entail direct measurement of the thermal and moisture expansion of treated and untreated specimens.

This approach was embodied in the early thinking of the RILEM 25 PEM Working Group (Réunion Internationale des Laboratoires d'Essais et de Recherche sur les Matériaux et les Constructions), which met between 1973 and 1978 to establish agreed testing methods for stone. Every test method put forward was considered in the light of questions like "Is this a useful test for measuring the extent to which a treatment strengthens stone?" or "Is this a useful test for measuring the extent to which a treatment immobilizes salts?" In the later stages of the group's work, the purpose of the framework became blurred by the introduction of the all-embracing questions, "Is this a useful test for assessing the durability of treated stone?" The individual questions had been intended originally to build up an overall picture of a treatment's effectiveness, and the late introduction of a specific question on durability tended to confuse this intent.

The approach is not perfect, not least because it tends to oversimplify decay mechanisms and to treat them in isolation. Nevertheless, it demands a thoughtful analysis of the functions that a preservative is intended to perform in any given situation, and this alone has much to commend it.

PERFORMANCE CRITERIA

The approach of the preceding section may be given a further degree of sophistication by establishing performance criteria for stone preservatives. Although the concept of performance specification has received considerable attention from the construction industry in recent years, its application to stone preservation appears to have been limited to studies by Gauri and by Sleater (see Bibliography under Performance Criteria).

A performance specification is essentially a statement of what a product is required to do, as opposed to a prescriptive specification, which defines the product in physical, chemical, and geometric terms. In its most elementary form, a performance specification for a stone preservative might read, "The preservative shall lengthen the useful life of stone without affecting its appearance." Such a specification would have no practical value, and a performance specification thus evolves into a set of more specific criteria, such as "Treated calcite specimens, exposed to a water-vapor-saturated 10.8 ppm SO_2 dynamic atmosphere, shall not show more than half the reaction of similarly exposed untreated specimens of the same material . . ." or "The surface of the preserved stone should not spall from the stone, decompose, change in appearance, or otherwise deteriorate when exposed to 1200 test cycles of water condensation/evaporation."

Such criteria are based on the approach of the preceding section but go one stage further by laying down quantitative limits for the test results. They have an air of arbitrariness, which is inevitable in view of the newness of the approach, but further usage should lend weight to the limits prescribed or else provide evidence to support amended values.

It would be tempting to assume that compliance with a comprehensive set of performance criteria would provide complete assurance that a treatment was effective and risk-free. But how comprehensive must "comprehensive" be? No practicable specification can possibly cover every aspect of performance, and loopholes must inevitably exist. It is particularly difficult to ensure that the test requirements are adequate when one does not know the chemical nature of every possible preservative. And it is almost certain that no product could be found that would meet all the criteria; certainly Sleater found no treatment, out of 56 tested, that met his requirements. Nevertheless, as he concluded, the assessment was valuable as a preliminary screening of materials for field testing.

The wheel has turned full circle, from field testing in an earlier section, through performance testing, and back to field testing. The next section examines some of the implications for both conservators and conservation scientists.

DISCUSSION

The preceding sections have given credence to the view that field testing, despite its limitations, is ultimately the only valid test for a stone preservative. Laboratory testing is useful for eliminating treatments that are unlikely to be effective, but at the present state of knowledge laboratory testing alone is insufficient. This is of little comfort to the conservator faced with pressing practical problems, to whom most of this section is devoted. For the researcher it means that attention must be given to the development of improved monitoring procedures for site trials.

Equally important is the need for detailed recording of site trials and for widespread coordination of results. It is easy for the researcher to publicize the results of trials in which the treatment is proved effective. It is less attractive to him, but no less important, to publicize trials in which disappointing results are obtained. There has been a tendency in the past to let unsuccessful trials fade quietly into obscurity, perhaps to avoid embarrassment. In order to build up a picture of what consitutes an effective preservative, however, it is essential that knowledge of all site trials be widespread and detailed.

But what advice can one give to the conservator, to whom the entire concept of treatment evaluation may seem irrelevant? "The stone is in such appalling condition that whatever I do cannot make it worse" is an argument often used to justify the use of an unproven treatment. (In fact, the argument is overworked and sometimes serves as a salve to the conservator's conscience.) The conservator must be given guidance on the scope and limitations of various preservatives, and it is up to the scientist to provide the technical element of that guidance.

In formulating conservation policy, a number of factors must be taken into account. First, stone preservatives are no substitute for common sense. In the case of external stonework, attention must be given to protection against rain, perhaps through the repair of protective features such as canopies or cornices or through the repair of open joints in masonry above. In extreme cases it may be appropriate to move the object in question to a controlled indoor environment.

Second, the use of stone preservatives may not be necessary if the stonework is easily accessible to a skilled work force. In such cases it may be preferable to maintain the stonework in sound condition by using adhesives or dowels to secure pieces in danger of falling away, by filling cracks with suitable mortar, by securing loose flakes with a judicious packing of mortar, and by filling water traps. Work of this nature can be undertaken whenever it is seen to be necessary, although it will not be sufficient if the stone is in very friable condition.

Third, it must be accepted that stone preservatives, like medicines, do not offer unending life. The time inevitably will come when the stone begins to deteriorate again, and retreatment may become necessary. No preservative should ever be used that precludes the possibility of eventual retreatment.

In deciding whether or not to use a preservative in a particular instance, it may be helpful to draw a distinction between stonework that is exceptionally valuable, either artistically or historically, and stonework that is of a more routine character. In the latter case the conservator may reasonably proceed with the application of a preservative whose record in field and laboratory trials so far has been satisfactory. In the case of valuable stonework, it is better for the conservator to assess whether the object's rate of decay is so great that unacceptable losses will take place within the next few years. If it is, the use of a preservative with a good record to date could not be regarded as irresponsible. If, on the other hand, the decay is reasonably slow, it is safer to leave the object untreated for the time being and to wait for experience to accumulate on other objects.

This paper inevitably has entailed presenting a lot of uncertainities, but one certainty can be cited. When asked whether one can be sure that a particular preservative will be effective and cause no harm, the answer is a categorical "no." However, it is hoped that this paper will stimulate further work, so that a rider may eventually be added: "No, but the risk of ineffectiveness and harm is so low that the treatment can be used without anxiety."

BIBLIOGRAPHY

[This bibliography is not intended to be comprehensive. Its purpose is to cite a few papers that contain typical examples of the various evaluation categories considered above.]

Evaluation by Natural Weathering

Clarke, B.L., and J. Ashurst. *Stone Preservation Experiments.* Building Research Establishment: Watford, England, 1972.

Moncrieff, A. The treatment of deteriorating stone with silicone resins: Interim report. *Studies in Conservation* 21:179 (1976).

Tests Primarily for Untreated Stone

Arnold, L., and C.A. Price. The laboratory assessment of stone preservatives. Proc. Int. Symposium on the Conservation of Stone, Bologna, 1975, 695.

Cormerois, R. Traitments préventif et curatif des structures. Proc. Int. Symposium on the Conservation of Stone, Bologna, 1975, 6.5.

Marschner, H. Application of salt crystallization test to impregnated stones. Proc. RILEM/UNESCO Symposium on Deterioration and Protection of Stone Monuments, Paris, 1976, 3.4.

Characterization of Treatments

deWitte, E., P. Huget, and P. Van Den Broeck. A comparative study of three consolidation methods on limestone. *Studies in Conservation* 22:190 (1977).

Accelerated Aging Chambers

Cooke, R.U. Laboratory simulation of salt weathering processes in arid environments. *Earth Surface Processes* 4:347 (1979).

Furlan, V., and F. Girardet. Méthode d'essai de vieillissement accéléré pour l'étude des traitements des pierres. Proc. Int. Symposium on the Conservation of Stone, Bologna, 1975, 713.

Sleater, G. A laboratory test programme for stone preservatives. Proc. RILEM/UNESCO Symposium on Deterioration and Protection of Stone Monuments, Paris, 1976, 6.13.

Tests for Assessing How Well Treatments Meet Objectives

Arnold, L., D.B. Honeyborne, and C.A. Price. Conservation of natural stone. *Chemistry and Industry* 17:345 (1976).

Gauri, K.L., et al. Reactivity of treated and untreated marble specimens in an SO_2 atmosphere. *Studies in Conservation* 18:25 (1973).

RILEM Working Group 25 PEM. Recommended tests to measure the deterioration of stone and to assess the effectiveness of treatment methods. Proc. Int. Symposium on Deterioration and Protection of Stone Monuments, Vol. 5. Paris, 1976.

Performance Criteria

Gauri, K.L., J.A. Gwin, and R.K. Popli. Performance criteria for stone treatment. Proc. 2nd Int. Symposium on the Deterioration of Building Stones, Athens, 1976, 143.

Sleater, G.A. Stone preservatives: Methods of laboratory testing and preliminary performance criteria. Technical Note 941. National Bureau of Standards: Washington, D.C., 1977.

Policy for Use of Stone Preservatives

Jedrzejewska, H. Some comments on ethics in conservation of stone objects. Proc. RILEM/ UNESCO Int. Symposium on Deterioration and Protection of Stone Monuments, Paris, 1978, 7.13.

Price, C.A. Brethane stone preservative. Building Research Establishment current paper, in press.

Torraca, G. Treatment of stone in monuments: A review of principles and processes. Proc. Int. Symposium on Conservation of Stone, Bologna, 1975, 297.

APPENDIX:
Proposal for a CCMS Pilot Study on the Conservation/ Restoration of Monuments

The term "monument" as used in this document includes all structures—together with their settings and pertinent fixtures and contents—that are of value from the historical, artistic, architectural, scientific, or ethnological point of view. This definition includes works of monumental sculpture and painting, elements or structures of an archeological nature, inscriptions, cave dwellings, and combinations of such features.

I. INTRODUCTION

Historic and artistic monuments represent the single most visible aspect of our history and culture. These monuments, mostly of stone construction, are universally threatened by the effects of pollution, urbanization, and public access, as well as by weathering cycles and other natural phenomena. Though there is national and international activity in the preservation of individual monuments, there is obvious need for increased cooperation among all those concerned with the development and implementation of national preservation plans.

II. ROLE OF CCMS

CCMS countries have a great concentration of stone monuments of varying degrees of antiquity, from the caves of Lascaux to the great monuments of ancient Greece and Rome, medieval cathedrals, renaissance architecture, and ultimately buildings of our century. Many of these threatened monuments have been treated in various CCMS countries under conditions of limited control. With its ability to facilitate international collaboration, CCMS could serve as a mechanism for developing uniform methods of reporting, for coordinating national research efforts, and for developing recommendations for treatment of monuments. These efforts would lead to the establishment of a general approach and would not be concerned with their application to specific monuments.

III. OVERALL PURPOSE

A. To enhance participating nations' abilities to minimize adverse environmental effects on monuments.

B. To develop options for governmental action to enhance conservation/restoration programs.

C. To serve as a model for international cooperation in the preservation of cultural property.

IV. STATE OF THE WORK

There are conservation and restoration projects under way at the local and national level in practically every country. Private institutions and museums are also involved in this type of work. Excluding the problems posed by purposeful human actions such as demolition, land development, and vandalism, monuments of value are most often damaged or lost due to the effects of air pollution, undermining and subsidence, and excessive vibration. In many cases, actions to restore or clean monuments have proved to have deleterious side effects.

After conducting the research necessary to diagnose the causes of deterioration, the techniques of conservation/restoration include:

A. *In situ protection* This covers methods that attempt to protect the monument in place. This is attempted through surface treatments or enclosures that seek to protect the monument from pollution or other harmful agents.

B. *Reduction or elimination of causes of damage* In some situations it may be possible to take such measures as reducing sulfur emissions or rerouting heavy traffic, rather than deal only with the resulting damage.

C. *Relocation* In some cases it may be necessary to remove the monument or certain elements of the structure from the original location to a site where they can be protected.

D. *Periodic restoration* Where other methods cannot be applied, it may be necessary to accept damage and to deal with it by periodically restoring the monument.

V. PROVISIONAL STUDY OBJECTIVES

The priorities in this list will be set at the experts' meeting:

A. To survey the state of the art of conservation/restoration methods.

B. To identify and develop uniform methods for evaluating treatments for such things as appearance, effectiveness, stability, reversibility, safety to associated building components, safety to workers, and economic feasibility.

C. To develop methods for the measurement of rates of deterioration and to determine mechanisms of deterioration leading to an understanding of the physical, chemical, and biological processes involved.

D. To develop recommendations for the maintenance of monuments. Within the context of the overall treatment these maintenance procedures could include repointing, cleaning, protection against biological attack, and isolation from groundwater.

E. If appropriate, the participating countries may undertake the study in depth of archetypal monuments as demonstration projects.

VI. WORK PLAN

The above activities fall into two types of study: research and documentation. The documentation projects should be completed by the end of the three-year period for the pilot study. The research initiatives should continue under the sponsorship of the individual participating nations. The uniform methods of reporting and cooperation developed in the course of the project may assure the successful coordination of these long-term projects.

A. *Analysis and Recommendations for Action* Under this task, lead countries would analyze the information collected as a result of the conferences. Recommendations would be prepared, presumably including guidelines for national programs in conservation/restoration. Possible demonstrations would be identified. Long-term projects such as the following would be initiated:

1. The design of suitable air pollution monitoring equipment and networks.

2. The design of a series of test walls in representative locations. These test walls composed of various treated and untreated stones would be exposed to typical environments:

    a. to develop criteria for monitoring deterioration of the test walls;

    b. to correlate the physical and chemical changes with air pollution data;

    c. to correlate data from such in situ testing with laboratory data.

B. *Documentation* Under this task a set of meetings would be convened to develop:

1. A census of treated monuments to permit an objective evaluation of empirical and scientifically controlled treatments.

2. An archive of literature on treatments and materials used in treatments.

3. The exchange of information on continuing research and treatment projects.

4. Uniform methods of reporting treatment procedures and materials.

5. Methods of regular reporting of the microclimates for individual monuments.

6. Models of monument microclimates.

7. Uniform methods for reporting the conditions of monuments by using techniques such as measured drawings, photogrammetry, and holography.

VII. RESOURCE ORGANIZATIONS

In carrying out the study, experts will work with data and material available from appropriate national organizations, both public and private, as well as such international organizations as:

    A. International Council of Monuments and Sites (ICOMOS)

    B. International Council of Museums (ICOM)

    C. International Centre for the Study of the Preservation and Restoration of Cultural Property, Rome (ICCROM)

    D. United Nations Educational, Scientific, and Cultural Organization (UNESCO)

    E. Réunion Internationale des Laboratoires d'Essais et de Recherche sur les Matériaux et les Constructions (RILEM)

F. International Engineering and Geological Societies.

G. Comité de la Pierre (ICOM, ICOMOS, ICCROM, RILEM)

VIII. PARTICIPATION BY NON-NATO NATIONS

In view of the universality of the problem to be dealt with in this study, appropriate non-NATO nations would be invited to participate, provided permission were granted by way of the usual procedure.

IX. PROJECT AS MODEL

Many problems in the preservation of our cultural heritage would benefit greatly from the CCMS approach to international cooperation. The successful implementation of the Pilot Study on the Conservation/ Restoration of Monuments could serve as a model for similar projects dealing with such problems as:

A. The preservation, in situ, of stained glass.

B. The improvement of museum and historical site security.

C. The protection of archeological sites.

D. The development of standards for the museum environment.

E. The development of standards for the protection of works of art in exchange programs.

F. The preservation of library and archival materials.

G. The preservation of ethnographic materials.

Index